SHAPING OUR ENVIRONMENTAL CONSCIENCE

SHAPING OUR ENVIRONMENTAL CONSCIENCE

✦

GARY COCHRAN

OLD RUGGED CROSS PRESS

Biblical references are quoted from the
following Bibles:
Living Bible (LB)
New International Version (NIV)
New American Standard Bible (NASB)

TABLE OF CONTENTS ✦

Chapter One

✦

The Problem

"Three pounds," you announce proudly, unhooking your son's bass from the pocket scale you always carry in your tackle box. "Bigger than my first one."

Your grin fuels his even more.

"Let's keep him, Dad," he says excitedly. "He'll feed all of us. Mom likes fish, too."

Now you have done it. He wants to keep his first good catch just like you did. But you glance around at the homes surrounding the pond. You note the neatly clipped lawns— fertilized, sprayed, watered. Then you look down at the storm drain your son stood on to fight the fish. A brown trickle pushes out from the pipe before succumbing to the green water in the pond.

"I don't know, Son."

Your grin fades, and so does his.

"Why not?" he asks pleadingly.

Rain forests make headlines. But the quality of our own little part of the world affects us more intimately. We live there.

Depending on where we live, we may face oil spills, or

overstuffed landfills, or endangered species, or wetland destruction, or hazardous wastes. We may face all these problems and more.

Are these headlines separate diseases infecting the body of human environment, or are they symptoms of a singular sickness? Each of these issues draws a cry of alarm throughout the world. Too often, though, the call to arms ends where it began: with more oil spills, more landfills closed, more endangered species . . . . And so it goes.

In a system as complex as our environment and with a creature as diverse as humanity, can there be a common root of infection erupting into the myriad environmental complaints we see across the globe? If so, does there exist a common cure? Many people ignore the common thread woven through the environmental fabric. But it is there.

As the Bible warns us, the love of money is the root of all kinds of evil. Not that money is moral or immoral. It is neither. The craving for wealth, however, tempts us to manipulate people and resources in a mad drive to find security and happiness in money. Rarely will anyone allow a bald eagle's nest or a redwood forest to block their path to riches. For most of us, it seems, only the threat of punishment motivates us to protect our environment, and then only to a minimum standard. This is not to say anyone desires a polluted environment. No one does. But environmental damage occurs as a side effect of the pursuit of wealth.

Development in coastal wetlands and estuaries, for instance, has created beautiful homes and productive businesses. Jobs have been added in local economies and tax structures have been strengthened. Many thousands of people have achieved their dream of living at the beach or on the bay. Developers and realtors have become rich through building on and selling former wetlands. We can say much in favor of wetland development.

But there is a trade-off. Ecology applies not only to biological systems but to economic systems, as well. As the world shrinks conceptually, we face squarely the truth of the old idiom, "there is no such thing as a free lunch." For each improvement in a former marsh, a net loss occurs. Often this loss extends far beyond the filled wetland.

Mangrove estuaries and salt marshes provide critical nursery grounds for many species of important commercial and sport fish. Sea grass beds furnish important habitat for many of these same species and others. Yet, in the Tampa Bay, Florida, area alone, a 1986 study reported 44% of the original mangrove forests and salt marshes had been destroyed, and 81% of the sea grass beds had disappeared.[1] Following these losses, the scallop and oyster fisheries in Tampa Bay collapsed, and populations of bait shrimp, spotted sea trout, and redfish declined dramatically.[2] Eventually, following a complete ban on taking redfish, the State of Florida declared this species a sport fish, thus removing it from commercial fishing pressure.

But many people missed the point. As habitat declines, so decline the resources dependent on that habitat. And human enterprises centered around those resources also decline.

No one need stretch their imagination to connect coastal development with the love of money. At some point, someone decided mangrove forests and salt marshes would provide better habitat for people than for wildlife—with a little improvement. Most of the population in those areas agreed or just did not care, and a generation or two grew up accepting such development as normal. An environmental conscience was formed. This conscience accepted wetland loss as inevitable and even desirable. Why? Because wetland loss seemed to be humanity's gain.

Today we are learning the far-reaching effects of past environmental decisions. Storms ravage coastlines denuded of protective vegetation and hardened with seawalls. Nothing at these sites dissipates the wind-driven wave energy before it smashes into coastal structures. Everyone pays the price.

In the United States, coastal construction is not protected by natural beach dunes and coastal marshes, but by federally subsidized insurance. Such insurance makes the risk of loss more acceptable, thus promoting development of barrier islands and coastal wetlands. This insured development is making a lot of people rich at great cost to others. The answer, then, is to cease coastal development. Or is it?

Banning development along coastlines would eliminate much of the pollution, congestion, and habitat destruction coastal ecosystems presently suffer. But this solution forgets humans comprise an intimate part of the ecology of this planet. Moving everyone away from coastal areas would stress other ecosystems in many of the same ways development stresses shorelines. We have to live somewhere.

If the populations of Cuba, Japan, Hawaii, and Great Britain were dumped elsewhere, where would they go? This solution attempts to remove humans from a large portion of their environment and is doomed. It is simply impractical.

On the other hand, the historic attitude of encouraging coastal development attempts to elevate humans above the natural constraints of their environment. This plan, too, has proven a failure.

Consider the severe degradation of Long Island Sound. Since 1985, the Long Island Sound Study has identified five priority problems: hypoxia (low dissolved oxygen), pathogens, floatable debris, toxic contamination, and the health of marine resources as related to water quality.[3]

Here, as elsewhere, humans have foolishly chosen to live in conflict with their environment rather than in balance with it. And who suffers more from foolishness than the fool?

The extremes of no development and overdevelopment create an imbalance of preservation on one hand and utilization on the other. Each concept is raised to an end in itself, and the human element inherent in both cases is almost ignored.

The National Park Service in the United States shows us one extreme. For years this agency allowed wildfires in some of the natural areas of national parks to burn. The Park Service apparently believed this practice allowed natural forces that occurred prior to human encroachment to keep these areas truly natural. But the Park Service missed an important point. Humans are on the scene now, like it or not.

Land developers usually take an opposite view. To many

developers, natural systems are worthless apart from their potential for marketing. Thus, human values apart from money are given little attention. Both of these viewpoints miss the distinction between the basic concepts of conservation and preservation. These concepts appear almost synonymous, but, actually, may exclude one another. Conservation implies wise use of a resource. Preservation implies non-use. In some cases, preservation may be essential to the goal of environmental protection.

If the Florida panther is to be saved from extinction, for instance, a portion of its habitat must be preserved from much human intrusion and preserved quickly. Probably as few as fifty adults survived in Florida in 1991.[4] This cat's life history and habitat requirements dictate preservation as the preferred strategy. They do not adapt like pigeons.

By and large, though, a conservation ethic provides the motivation we need to manage our environment for our greatest long-term benefit. The Bible exhorts us to make the whole earth our home and to use the earth's resources as God-given gifts. Whether or not a person believes this offer, he or she surely must realize the need to conserve the resources of this planet to serve an increasing human population. Only this attitude will encourage people to live as an integral part of their environment rather than as its creator.

Here the National Park Service stumbled. Rather than conserving resources through sound management practices, the Park Service wasted resources through an idealized notion that conditions of past centuries should exist today. Wildfire consumed resources that might have been conserved.

This is not to say the Park Service should stress temporal utilization of park resources. But it should recognize that humans form part of the ecology of national parks. Management schemes then would integrate the human factor into the natural systems in these parks. In this way, resource utilization would become more than consumption. At the same time, protection and management tactics would conserve park resources for human benefit today and in the future.

But our concept of resource use will have to change first. Historically, resource utilization has resulted either in exploitation or conservation of the resource. Even today exploitation often reigns because money can be made more quickly by exploiting a resource. Swamps generally produce minimal income, for instance. Filled, however, swamps transform into home sites and power plants.

Aside from the immediate environmental loss we experience exploiting a resource, we develop a conscience that accepts exploitation as normal. Earth's rich productivity and the historical perception of unlimited natural resources helped ingrain such an environmental conscience in humans. Many generations have lived with the idea the wilderness must be tamed to be valuable. We have etched this apparent conflict between ecology and economy into the stone of our hearts. And some hearts are stony, indeed.

Many people think of their hearts merely as organs. They do not accept the sentimentality of "heart" applying to the vital center of their being. However, human life reaches deeper than the mind. We have a conscience, good or bad.

Our conscience determines our conduct toward the environment just as it determines our conduct toward other people. Some people abuse others. Some people pollute the environment. Our conscience can justify all manner of abusive behavior, disrespect for authority, and contempt for prudence. It can cloud our thinking, too.

Americans, in particular, link pollution to industry, developers, and government inaction. But people pollute. People run industries and build developments. People use confusing language and meaningless statistics to convince others contaminated emissions do not pollute. And people can be gullible. Consequently, the public often does not recognize a serious pollution threat, and government regulators focus their attention on other matters. They all get lost in a word game.

Even so, modern society thrives on definitions. Intuitive

knowledge has value only when it is supported by facts and figures. Perspective often determines the "facts." Thus, a cup containing half its volume may look half full or half empty depending on the observer. An ecologist may see anything that disrupts the natural order as pollution. A developer may see anything in its natural state as a waste. A government regulator may see anything not prohibited as being acceptable.

Such perspectives relate to our conscience. If an industry convinces people its emissions do not pollute, it scores a victory for the industry. If a developer successfully skirts the rules, others applaud his or her creativity. If the government fills regulations with vague language and toothless stipulations, it may appease the public *and* the polluters. And an environmental conscience is enforced.

Our conscience helps us choose the most appropriate use of a resource. Years ago, people decided air and water made good waste baskets. This attitude has moderated, and few people today advocate dumping raw sewage in a stream. But much of our environmental "enlightenment" is actually camouflage.

For instance, engineering plans disguise wetland destruction by implying mitigation replaces lost wetlands. Such plans ignore evidence showing the difficulty of creating a wetland that functions as well as the natural system it replaces.

In addition, vast numbers of people presume themselves to be the earth's only creatures of worth. Thus, they justify pollution as long as they experience no detrimental effects themselves. This presumption hampers a farsighted environmental conscience that understands we are spoiling our home.

It seems few people realize the world around them is a forecaster of their own health. Can we reduce the fisheries productivity of marshes and estuaries and not feel the pinch at the grocery store? Can we print an overabundance of newspapers and expect them not to litter the roadsides and pack landfills? Can we expect medicine to conquer bronchitis while we smoke and pollute the air?

A misguided environmental conscience may ultimately lead

to resource depletion and irreversible loss. Many extinct plant and animal species plainly illustrate this fact. However, we have not learned our lesson.

We cannot overharvest grouper and destroy their habitat and still expect this important fishery to survive. We cannot let agricultural soils erode away while we farm them intensively year after year and still expect their yields to grow. We cannot discharge nutrient-laden effluents into lakes and bays and not expect algae growth to explode.

The time has come to pay attention to the environment. We must live here all our lives, and future generations depend on our environmental integrity. We can enjoy living here. At the same time, we can manage the environment to provide for our needs and the needs of our children and grandchildren. We need only control the greed that tears at our conscience and tempts us to exploit our world for selfish gain. To do this, we may find we need a drastic change in our very nature.

By nature, we work to improve what already exists. Not that improvements are necessarily bad. They can be quite good, in fact. The problem arises when we disregard the function of an existing natural system in our zeal to make it more immediately useful. That is how we turn a profit.

Profiting from our endeavors is not wrong. Most people work at some job to make a living, and they should. And workers have every right to expect appropriate compensation for their labor. Many people, however, are not content with their wages.

People everywhere race to accumulate wealth and learn to justify all manner of immoral behavior in their quest for money. They kill, steal, defraud, threaten, and do all kinds of evil things. And they exploit resources created for eternity to build temporary wealth for a relative few. They also become blind. They do not see that exploitation is the natural result of the lust for money.

Ironically, value systems cause much of the blindness. Industries, individuals, public utilities, and others view natural resources from different, and often conflicting, perspectives.

A homeowner may see a great river as an asset to his or her property value. A power company may view that same river as a convenient source of cooling water for its plant. A municipality may consider the river a handy place to dispose of treated waste water. Who is right or wrong?

In the United States, personal freedom provokes the notion that ownership magnifies our rights. Americans figure owning river front property grants them the right to alter that river front in any manner they choose. They forget, or ignore, that what they do may affect someone else.

People even kill one another with their thoughtlessness. The United States Environmental Protection Agency added environmental tobacco smoke as tenth in its list of human carcinogens, blaming it for 3,000 lung cancer deaths per year in the United States.[5] No one needs to worry about cigarette smoke hurting a tree or a squirrel. But we all need to insist a smoker's personal freedom does not endanger someone else's health.

Even environmentalists can overlook such common pollution in their zeal to discover some new and exciting issue. Many Americans, however, suspect cigarette smoke threatens more people than does fallout from asbestos building materials. We possess a wealth of data proving the health risks from cigarette smoke. Yet, we continue to pollute with cigarettes. We even subsidize tobacco farming. Where is the sense in all that?

Perhaps not enough nonsmokers recognize the connection between environmental health and personal health. Perhaps we all just want to determine our own destinies. However, if we do not exercise foresight, personal freedoms will not mean a thing because the freedom to choose will disappear. We cannot choose to protect passenger pigeons. They have already vanished. And a person dying of lung cancer will soon lose all choices.

Individuals are not alone in their lack of foresight. Industries and businesses must compete in the marketplace and they are sorely tempted to grasp what they need at the expense of others. No matter who we are, a selfish motive precludes our considering the needs or desires of others. It also makes our blindness grow darker. We reject the notion we sometimes should defer our individual rights for the good of others.

A particularly narrow perspective is displayed by the oil

and gas industry. A threat to a future oil supply initiates an immediate price hike by oil companies. The same threat also gives them ammunition to lobby for permission to drill in untapped areas of the Florida and California coasts. The oil companies want more profits. But many Floridians and Californians want to protect their coastal resources and the economic production of those resources.

Again, that word: "economics." Oil wells spell prosperity to oil companies and those providing goods and services to them. Oil wells generate tax revenue. But drilling in coastal areas is potentially devastating.

In the eastern Gulf of Mexico, near-shore waters are shallow. The islands there support a thriving tourist industry, and the Gulf waters and inshore bays support important commercial and sport fisheries. The estuaries provide indispensable nursery grounds for as much as 90% of the commercial and sport fish and shellfish of the region. Consequently, the livelihood of many Floridians depends on clean beaches and productive coastal waters. One oil spill could wreck the livelihood of a great number of people. In this case, conservation provides public benefit right now.

Conservation requires a number of considerations, among which are the nature of the resource and its wisest use. Natural resources are basically renewable or nonrenewable. Nonrenewable resources have a finite supply which will be exhausted in time. Petroleum is a nonrenewable resource. Renewable resources, on the other hand, can be managed to promote an indefinite supply. Trees are renewable.

Conserving these two types of resources takes different strategies. Wise use of a nonrenewable resource suggests we carefully consider the speed at which we use that resource and develop alternatives to its use. We may decide we have enough to last a couple hundred years and do not need to worry about alternatives now. But the decision exposes a poor environmental conscience. Such decisions also handicap future generations by leaving them with an insufficient supply. Poor planning today creates crises tomorrow. We see this result clearly in congested cities.

Renewable resources also demand responsible planning. But planning, in this case, aims at perpetuating a resource for indefinite use, even if that "use" is the pleasure of knowing a rare species still exists. In any case, as the human population increases, even renewable resources will have less of the earth's surface on which to be renewed. To conserve a renewable resource, we still may need to develop alternatives to reduce demands on any one resource.

Thus, conservation requires hard choices. Priorities need to be established and previous priorities scrapped if proved wrong. We may even have to choose between resources.

No considerate person welcomes the extinction of a plant or animal, at least not one considered desirable. Yet, we must face the fact, unwelcome as it is, that we may not be able to maintain a species throughout its original range and still meet human needs.

The key is need. We need clean air and water. We need adequate shelter and wholesome food. We even need entertaining and relaxing diversions. But we do not need automobiles that cruise at 100 miles per hour or pleasure boats that go sixty. We do not need homes in two different states just because we dislike the climates of both. People can adapt.

However, many plants and animals cannot adapt well to people. Therefore, bald eagles may disappear from one region and be maintained in another while attaining our goal of resource conservation. When this occurs, we say a species has been extirpated from a particular region (but has not become extinct). Here, though, we must be careful. Extirpation eventually may reach the point where a species cannot reproduce in sufficient numbers to survive. So conservation returns to the trough of hard choices and serious priorities.

Setting conservation priorities means we must consider the type and level of use of a resource. One resource may support multiple uses while another resource supports only one. Forests can support multiple uses, though not always on every acre. The same fact applies to water resources. A mining area, however, may not permit any other use until mining is completed and the area reclaimed.

But environmental concerns reach beyond such issues as conservation practices. Environmental quality affects human health. Everyone wants to breathe clean air and drink clean water. We know our health depends on it. But environmental quality includes much more than clean air and water and affects our lives in ways sometimes hard to identify.

In 1990, *Resources* magazine reported the infant mortality rate in the Aral Sea area of the former Soviet Union reached 250 out of 1,000 compared to a countrywide rate of 22 out of 1,000. The average life expectancy in this region was reported to be 22 years less than the average in the Western world. Environmental causes turned out to be the culprit. The Soviets' solution: the "pay to pollute" principle. The government determined a payment amount for allowable limits of emissions. Then three to four times that amount was charged generators that exceeded the allowable limits.[6]

This tactic is questionable at best. But it was also reported the Soviet Union had adopted laws and regulations complex enough to achieve environmental standards equal to United States standards by the year 2005. The upheaval in the former Soviet Union will undoubtedly affect the implementation of these standards. Such standards, though, may become reality as these countries move closer to a market-based economy. Nevertheless, we all need to remember the Soviet experience. Totalitarian governments do not demonstrate environmental integrity.

Even capitalist societies stumble in their attitude toward the environment. An auto parts company sold a book entitled *How to Bypass Emission Controls—For Better Mileage and Performance*.[7] Apparently, this company believed short-term fuel economy was more important than long-term air quality.

Some people would argue fuel economy yields greater benefits than emissions control. Others would argue the opposite. Regardless of which position we choose (or another one), we need to guard against allowing money alone to dictate our priority. Otherwise, we will never escape the habit of looking at our lives, and the lives of others, through the clouded lenses of self-interest. More is at stake than immediate profit.

If we can divorce ourselves from this tendency to base decisions solely on money, we can consider more alternatives for achieving true resource conservation. In the process, our environmental conscience will begin to include ideas and alternatives perhaps new to us.

At one time, people thought the universe centered around the earth. Eventually, we learned the earth is a small component of one solar system among many scattered throughout an immeasurable universe. But this knowledge was hard won. People just did not want to believe it.

Today we know the world is finite. The resources on earth are all we have. But this knowledge has failed to produce a humility of spirit that encourages wise use of our world. Instead, our arrogance insists we will design away flaws and shortfalls in our natural environment and replace them with technological marvels. This conscience deludes us into believing we can live beyond the constraints of our environment. In our humanistic zeal, we forget God alone can create something out of nothing. Our conscience needs to change.

We are blessed because our conscience is dynamic. We learn values: then our decisions and behavior both reflect and enforce those values.

Some people oppose the teaching of moral values. They say it represses a person's identity and imposes artificial restrictions on a person's life. But we should thank God values can be taught. Order and peace in society demand moral values.

Our conscience can be shaped and, in turn, can shape our actions for good. We need not be pawns of expediency. Even children can learn to judge right and wrong, good and bad, responsibility and capriciousness. It just takes time.

Tactics to help shape a sensitive environmental conscience are presented later in this book. For now, remember each of us can shape his or her own conscience, and, by example, help shape the conscience of someone else. Our future depends on it.

NOTES TO CHAPTER ONE

1. *Habitat Restoration Study for the Tampa Bay Region* (St. Petersburg, Florida: Tampa Bay Regional Planning Council, 1986), p. 5.
2. Ibid.
3. "Long Island Sound Study Proposes Interim Actions to Combat Hypoxia," *Coastlines* 1, No. 3 (Dec.–Jan. 1990-1991): 5.
4. David S. Maehr and Chris Belden, "The Endangered Florida Panther," *Florida Wildlife* 45, No. 3 (May–June 1991): 16.
5. "Worth Repeating," *Resources* 12, No. 5 (Nov. 1990): 16.
6. "News Front," *Resources* 12, No. 5 (Nov. 1990): 2.
7. P. Aarne Vesilind and J. Jeffrey Peirce, *Environmental Pollution and Control*, 2nd ed. (Ann Arbor, Michigan: Ann Arbor Science Publishers, 1983), p. 371.

CHAPTER TWO

✦

CONSCIENCE SHAPES LIFE

The Bible suggests our conscience largely shapes our quality of life.

Everyone possesses an environmental conscience that influences practically every decision they make. That may seem like a strange claim. But our environment extends far beyond the natural systems surrounding us to include virtually every aspect of our world, human and otherwise. Thus, our attitude toward our world both influences and is influenced by our environmental conscience. And we can expect tradeoffs with every environmental decision we make.

Usually, one side of the issue is money. The question then becomes one of priorities. Is conserving the resource more important than maximizing profit, or vice versa? It seems profit normally wins over the resource yielding that profit—not always a laudable practice, but understandable. Few companies, for example, could exist for long without showing a profit year in and year out.

Still, looking at a natural resource only from an economic perspective blinds us to less tangible benefits of that resource, yet benefits that may be equally important to our quality of life. Modern forestry in the United States provides a good example. Forests often are manipulated through their total harvest, planting in a single species, and harvesting again when the planted trees reach a predetermined size. This monoculture cycle has greatly increased marketable wood production over what would have occurred naturally. But wildlife production has sometimes suffered.

A landowner who wants to produce game, or even nongame, animals together with wood may need to alter his or her silvicultural strategy. Planted stands should be smaller and be interspersed with natural forest stands and open areas, even cultivated or fallow fields. Diversity of land use increases the "edge" effect, thereby producing more food and cover. But this same diversity can increase the cost of wood production.

No doubt people will enjoy more hunting, bird watching, and wildlife photography. However, a dollar value can be difficult to assess for land uses such as these. Hunting leases may provide some income. But these uses of the environment usually generate less tangible benefits than selling a crop, yet benefits that improve the enjoyment of life. Some of these values may even enhance the future earning potential of the property.

For most of us, such intangible or passive uses require that we alter our environmental conscience. We must subjugate our profit motive to allow for a less restrictive perspective. The resource itself contains an intrinsic value not dependent on what can be marketed from that resource. Unfortunately, some people do not enjoy the luxury of choosing environmental alternatives. They struggle just to eat.

An economy/environment conflict has raged in the Philippines for years. The Philippines is blessed with great ocean reefs, and these coral reefs harbor a bounty of tropical fish popular in the aquarium hobby. Many Filipino fishermen earn a meager living capturing these fish for export. These fishermen have learned cyanide gas will stun their prey and increase their harvesting efficiency and, thus, their income—an understandable attitude when we realize the poverty in which these fishermen try to raise a family. But cyanide can kill as well as paralyze. The aquarium press has reported for years about the reef destruction partially resulting from cyanide poisoning. Also, some exporters have stopped buying cyanide-captured fish. Still, this method of capture continues. These fishermen's poverty discourages long-range conservation. They must survive today. A destructive environmental conscience has formed out of the commendable desire to meet family needs.

The tradeoff has not always been recognized. How many generations can live off these reefs if cyanide poisoning continues to be added to the other stresses these reefs must endure? This is one instance where less efficiency, netting, may provide greater benefit by protecting future earning potential. Improved methods have increased the efficiency of netting, and the use of cyanide has been made illegal. Now only time will tell whether or not the environmental conscience of Filipino fishermen will change.

It would help if the market's conscience would change, too. As long as a market exists for cyanide-captured fish, fishermen will want to use cyanide. It is a matter of basic economics.

Tropical marine fish are expensive since most species are not yet raised commercially. Exporters pay less for cyanide captured fish because they expect a high death rate. Importers pay less for the same reason. So, a less expensive fish winds up at the retail store, and customers see a bargain rather than an environmental abuse.

This situation may be changing. The aquarium press has reported a preference for netted fish is growing among exporters, importers, and retail customers. Perhaps human conscience is finally shaping an economy for the good.

The Philippines is not unique in harboring a conservation conflict. Neither is the Twentieth Century. Conservation rules existed in biblical times. Mosaic Law commanded fields lie fallow every seventh year to allow a harvest for the poor and for the animals (ref. Ex. 23:10-11). At the same time, productivity of the land was restored. God promised the sixth year's yield would sustain the nation of Israel until the eighth year's crop was harvested (ref. Lev. 25:20-21).

This system succeeded while it lasted. Eventually, though, Israel got away from that practice and productivity declined. A tradeoff occurred. Short-term yield (a seventh year's crop) caused a long-term decline, and God finally removed the people from the land and restored the unobserved "sabbaths" to the land (ref. Lev. 26:34-35). Even for those who do not accept the biblical tradition, a biblical principle exists: our present activities affect our future.

In November 1990, *Resources* magazine reported about concerns that electric and magnetic fields (EMF) may cause cancer under certain circumstances. Electricity generates EMF. It hardly seems possible today to live without electricity, and who would want to? But studies have revealed an apparent correlation between proximity to EMF and cancer in children. A statistical relationship was shown between childhood leukemia cases and proximity to high-voltage power lines, perhaps as much as a one-and-one-half to two times greater risk of cancer. Additional concern exists about EMF from computers.[1]

Nevertheless, twenty years of study have failed to determine a direct cause-and-effect relationship between EMF and human health.[2] So, we just do not know. Our ignorance should prompt us to consider possible unwanted side effects of our dependence on electrical power. Not that we should start pulling down power lines. But we should proceed cautiously enough to protect ourselves from negative effects of a very positive feature of modern life.

We need to remember our environment extends beyond the natural world. Alterations to the environment, in turn, become part of the environment. The effects are not always desirable.

Too often human interaction with the environment has caused failures in spite of advances in technology. Crowded and unsanitary communities proliferated the rats and fleas that spread the great plagues throughout Europe during the Middle Ages. The Dust Bowl days in the United States point out the failure of shortsighted farming methods. The deterioration of a formerly great fishery in the Great Lakes flashes a warning about unregulated growth and environmental ignorance. Overdevelopment has coupled with an undependable supply of rain to bring about water-use restrictions in Florida and California.

The ways we have used our environment, rather than our presence in the environment, have caused or complicated the environmental problems we face now. Development carries a price.

Halting development, though, hardly provides for an increasing population on a finite planet. We must live somewhere, and we must eat. But we need to realize that the benefits from development may cost us dearly in terms of diminishing resources. One person's gain can be another person's loss.

Human benefits exist in developing and in not developing. Today we see more clearly our obligation to weigh these sometimes conflicting benefits and to consider alternatives on both sides. Still, the goal of a quick profit usually outweighs more protective considerations.

Remember the Philippines' experience of fishing with cyanide on its coral reefs. There, as elsewhere, poverty has blocked an enlightened environmental conscience. This should not surprise anyone. Economics drives us wherever we are.

Have you noticed environmental protection issues attract an economically elite leadership? Those who "have" tend to place a greater emphasis on a clean and healthy environment than those who "have not." We can hardly expect otherwise. California redwood forests probably mean little to a poor inner city family in New York City. These people suffer an environment of poverty, crime, sickness, and decay. Ecologists would do well to include this situation when addressing national and global environmental issues. Environmentalism arises from prosperity.

Anyone who wants to, however, can achieve a more sensitive environmental conscience. They can begin by broadening their view of what the environment really includes. Then they may better see the environmental ramifications of financial and technological decisions.

For instance, we can use water attractions to lure more people into a naturally droughty area. But by doing so, might we also exacerbate a water shortage? Or might we increase our dependence on automobiles, with their polluting emissions, through poor urban planning? Do we even block industries from building new, more environmentally friendly plants because of exorbitant permitting fees?

Environmentalists should address questions such as these just as much as the question of how to save Aleutian geese in Alaska. Had our society addressed such questions sooner, these geese might have a brighter future now.

A broad environmental viewpoint frees us to consider more than just money when choosing environmental options. Rather than arguing about how long a resource might last, we can agree usable supplies will run out. Then we will examine the wisest alternatives in conserving the resource. We will also explore alternative resources and ways to use less.

A broad viewpoint also helps us look beyond problems that are really symptoms of a greater sickness. Humanity decries, discusses, researches, and regulates pollution. Too often, though, we fail to see pollution as a by-product of greed and poor planning.

Because of this, many people think overpopulation is the ultimate cause of pollution. If that conclusion is true at all, it is true only on a limited scale. Humans have hardly filled the earth. Portions of the earth may be overpopulated because the worldwide distribution of resources has not favored those locales. People in more productive areas compound the problem because they resist sharing what they have with others. In addition, global politics often blocks the distribution of resources by those willing to share.

These problems point out the need for people worldwide to reshape their environmental conscience. Otherwise, we will keep viewing pollution itself as the "bad guy." We will overlook the real culprit. And we may overlook the cure.

Many people see technology as the solution to pollution. Indeed, advances in wastewater treatment, solid waste management, and water and air quality treatment have greatly reduced the negative impacts of human activity on the environment. But technology can pollute, too.

We produce hazardous materials faster and cheaper than we can safely dispose of their wastes. The public benefit of these materials may be well documented, but their public costs seldom are presented as strongly. These costs may not even be understood. The huge volume of these materials multiplies the problem of their disposal. To complicate matters, today people look backward and reminisce about the "good old days" but seldom recognize the pollution produced from past technological advances.

The steam engine, for instance, was perfected during the late Eighteenth Century. Thus, a revolutionary power source freed people from the need to be located around rivers that could supply water power. Cities developed as factories were built, and labor moved from rural areas to work in the factories. Railroads soon connected cities and reached into the sparsely populated countryside. In a way, the world shrank. Markets then expanded with the improved transportation and stimulated farm and factory production. Industrial improvements and new inventions accelerated the Industrial Revolution and launched humanity to high levels of technologically enhanced lifestyles.

We forget something, though. With the improvements came tradeoffs: oceans of wastewater effluent and sludge; tenements full of crime, disease, and despair; mountains of refuse concentrated in "sanitary" landfills; air pollutants that block the sun and generate public health warnings; water pollutants that stimulate vast algae blooms in lakes and streams; waste materials too dangerous to be handled by most of us. And much more. Universal familiarity with these problems hinders us from seeing that they resulted from a technological revolution no one wants to rescind.

We need not decry technology. But we need to recognize the advances that improve our health, increase our productivity, and strengthen our earning potential also affect our environment, oftentimes negatively. We would do well to examine both sides of the coin before we embrace an advancement as an improvement. Caution may add to the immediate cost of a product. Caution also may prevent a greater future cost to correct something gone wrong.

Caution also will reduce the shortcomings of considering environmental matters on a site-by-site, permit-by-permit basis. The ubiquitous problems of development can then be addressed through long-range planning and resource management. In coastal areas, in particular, population pressures illuminate the fallacy of shortsighted development policies.

About half of the United States' population now lives in coastal areas. That means nearly 110 million people are packed along the United States' coasts of the Atlantic and Pacific oceans, the Gulf of Mexico, and the Great Lakes. This population is projected to increase 60% by the year 2010.[3]

This situation should at least hint to us that many environmental problems of our coasts are tied to development patterns. The disruption of natural processes threatens the coasts' ecological and economic values. Consequently, an exploding population along the coasts can diminish the very qualities that attracted much of that population.

Does this sound far-fetched? By 1988, the population density of United States coastal areas had soared to over four times the national average, or 341 persons per square mile. Boston, Philadelphia, San Francisco, and New York City each exceeded 10,000 persons per square mile.[4]

The United States Department of Commerce considers population per shoreline mile to be one indicator of environmental stress. In 1988, the United States averaged 1,177 people per shoreline mile, and the projected figure for the year 2010 is 1,358. Illinois packs in nearly 92,000 people per shoreline mile.[5]

Development necessary to feed, house, employ, and entertain this dense a population obviously stresses coastal ecosystems. Planning for the resource needs of these people mandates caution.

Caution in resource planning will produce at least one other benefit. Caution will free us from the trap of resolving one environmental problem by damaging another environmental function. Environmental problems relate largely to ecosystems. A solution to one problem should avoid damage to another element in that ecosystem. Otherwise, we cannot practice effective environmental management.

Development many times has countered environmental management efforts. Such is the case with developing barrier islands. By nature, these islands form dynamic ecosystems, constantly moving, reshaping, washing out, filling in. But people want to live on them. Since development demands protection from financial loss, shorelines are hardened with sea walls and revetments. Beaches are manipulated to replace sand moved around by normal currents and storms. And taxes are levied to pay for these measures and to subsidize the insurance necessary for the islands' structures.

A more sensitive conscience might recognize that barrier islands evolved as the natural result of dynamic coastal forces. At the same time, sheltered bays developed behind these islands, creating unbelievably productive estuaries. These islands now form barriers between the open ocean and the mainland and attenuate the destructive forces of storms. A conscience that allows these islands to function within their ecological niche recognizes the immense value of barrier islands to far more people than those wealthy enough to live on them. But such a conscience may also lose money.

In Twentieth Century society, practically anything can be economically justified even if it is not environmentally sound. Attaching some kind of public benefit to the issue almost guarantees its acceptance. But public benefit deceives.

Oil and gas production, for instance, fuels a worldwide economy. Yet, the distribution of the raw product forces some nations to be dependent on others. Can the United States increase its domestic oil production to a level that immunizes the country against Middle East oil policies? It seems unlikely. Nevertheless, this argument promotes oil production in extremely sensitive natural areas, such as the Florida Keys. The public benefits, some say, outweigh the pollution risks. But do they? The impacts of a major oil spill can be dramatic, extensive, and expensive. The impacts also can be insidious. Ocean reefs and coastal zones are comprised of complex and sometimes fragile ecosystems. Reporters converge on an oil-soaked beach. But the slower cumulative effects of an oil spill may go unnoticed by the general public. A dying pelican makes a touching picture and a heart-rending story. The long-term decline of a reef or estuary is much more serious but much less obvious. Cameras cannot photograph snook and tarpon that are not there.

But fishermen have known for a long time an important key to fish production lies in the nursery grounds. Coastal estuaries provide critical habitat for juvenile stages of many commercial and sport fish and shellfish. Reducing the productive capacity of estuaries through habitat destruction or pollution reduces the fisheries output of these areas. Reduced production will be felt initially by fishermen through reduced catches, then by consumers through higher prices. Do not think what goes on in Florida does not affect people in Tennessee.

Wherein lies the public benefit illusion: where do we draw the line? It seems we draw the line at whatever point supports our point of view. When our point of view considers finances alone, we support only those practices that maximize profit or minimize cost. This point of view is not necessarily wrong. With it, though, we do not always think too far down the road. If we are to make sound environmental decisions that recognize long-term values, we must broaden our conscience beyond our own little world. We must study the small pieces of a big puzzle.

Ask yourself: is it possible to lure certain industries to states or countries with lax air and water quality standards? An industry there can spend less money on pollution control systems. This industry then may hire a large number of local employees and add to the area's economic health. Thus, a public benefit occurs.

But air and water pollution are rarely confined to a local problem. Air and water move between states and even between countries. Consequently, a locale with lax standards may enjoy an economic advantage at great cost to its neighbors. Now where is the public benefit?

Regardless of our attitude toward public benefit, we are not always helped by the rapid changes in the world today. To be sure, society has been blessed by technological advancements that have eased our lives, by medical advancements that have lengthened our lives, and by cultural advancements that have enriched our lives. But this same progress has added environmental stresses which sometimes have cost us dearly.

We continue to deposit the demands of an exploding human population at the doorstep of our natural world and further stress an already stressed environment. As our affluence increases, so do our demands. Truly, we should marvel at the resilience of our world.

Sometimes governments stack public benefits squarely in the way of private interests. Then conflict arises because a private enterprise is curtailed if not made impossible.

The United States government, for example, prohibits people from disturbing nesting bald eagles, and for good reason. Eagles may abandon a nest intruded upon by humans. In Florida, wildlife biologists advise property owners of ways to comply with federal guidelines. In some cases, compliance results in a buffer zone around an active nest. Human activity within this zone risks disturbing the nesting eagles. Such protection has been blamed for altering subdivision development and blocking road construction. Property owners, in some cases, have complained of reduced sale prices for their land.

Without aggressive protection, however, bald eagles probably would be extirpated in most of Florida. Some people would applaud that outcome. But many Floridians believe the eagles' disappearance would impoverish Florida even while generating a few dollars. Preservation, then, can be in the public's interest, too.

In addressing the question of public interest, someone must decide what the public benefit is and when it outweighs private interest—not an enviable task. That decision undoubtedly will face opposition. But people's environmental conscience can be shaped so they value a natural resource over maximized income. In earlier years, few Americans possessed that conscience, and they hunted Carolina parakeets to extinction. Americans are more affluent now and can deem a plant or animal more valuable than great profit. That is a good position to be in. It proves we can make money and conserve resources at the same time—if we are not greedy.

Mangroves portray clearly the conflict between private and public interests. "Mangrove" embraces a variety of tropical and subtropical evergreen trees growing along tidal shores. Mangroves grow widely in the tropics, but in the United States, they are confined primarily to the lower peninsula of Florida. They do more than look pretty. Mangroves in Florida perform functions vital to the ecology and economy of the Gulf of Mexico and southern Atlantic regions. These trees are so valuable, the State of Florida protects its mangroves for such reasons as the following (ref. Ch. 17-321, Florida Administrative Code):

1. Mangroves protect the shoreline against erosion.
2. Mangroves provide important habitat for a diverse community of plants and animals, including a number of endangered and threatened species.
3. Mangroves concentrate organic matter in coastal estuaries, creating a base for a complex food chain utilized by an immense variety of marine organisms.
4. Over 90% of Florida's commercial fish species depend on mangrove estuaries for nursery grounds.
5. A host of migratory birds find a dependable winter resting ground in Florida's mangrove estuaries.
6. Mangrove wetlands stabilize sediments, assimilate nutrients, and filter particulate matter, thus improving water quality.

Do these values induce people to protect mangroves? Sometimes. But mangroves grow in the same places many people want to build homes. The mangroves sometimes interfere with a clear view of the water and may harbor wildlife some people find objectionable, such as raccoons and opossums. In such cases, mangroves are often cut down. Judicious trimming can be justified in residential areas. But a study by Florida's Department of Natural Resources showed repeated trimming stresses mangroves to the point where their ecological function is sharply reduced. Shoreline erosion increases, seed production decreases, attractiveness to shore birds dwindles, and nutrient input to the ecosystem declines. This study confirmed conclusions of earlier studies that cutting harms the mangroves themselves, the estuarine environment, and the wildlife dependent upon mangroves for habitat.[6]

So, why would anyone cut down mangroves? Boaters cut mangroves to build docks. Public utilities cut mangroves to construct power lines and storm sewers. Surveyors cut mangroves to survey property. No problem. These alterations usually create minor impacts which recover quickly.

Most often, though, people cut mangroves because they want an unobstructed view of the water. To these folks, waterfront property ownership includes the right to see the water clearly. But the cumulative impacts from repeated severe trimming seriously diminish the ecological value of the mangroves, thus reducing their value to the rest of us. Most people who benefit from mangroves do not live on the water.

Rarely, though, will a waterfront owner consider such an altruistic notion. Fishermen, bird watchers, seafood lovers, and others benefit immensely from natural mangrove estuaries and suffer from mangrove loss. But a clear view of the water sells property. For waterfront owners to benefit others, they may have to restrict their own use of their property as well as the power of the water on its resale. Seldom do most of us yield to such a conscience.

Somewhere in the fight over mangroves lies a compromise. This compromise will protect the integrity of the mangrove ecosystem while allowing waterfront owners a view of the water. However, this compromise will not be found unless both sides cooperate. Waterfront owners and environmental regulators both need to "give." And everyone needs to consider the fruit of what they do.

The Bible cautions us about greed and selfishness. Society recognizes the ideal of selflessness and sometimes celebrates those who subjugate their own interests to help others. Yet, we somehow miss an important point. Generations ago King Solomon recognized the one who is greedy of gain troubles his own house (ref. Prov. 15:27). Solomon was no fool. The ones who suffer most from our self-indulgence are ourselves.

Until we learn to look at our environment through eyes unclouded by selfishness, we will never see the true value of our world. A resource treasure exists on Earth to yield a quality of life few of us have ever enjoyed, despite our personal wealth. The unwise use of our environment has robbed us all.

NOTES TO CHAPTER TWO

1. Susan Ainsworth, "A Current Affair," *Resources* 12, No. 5 (Nov. 1990): 3.
2. Ibid.
3. Thomas J. Culliton et al., *50 Years of Population Change Along the Nation's Coasts 1960-2010* (Rockville, Maryland: U.S. Department of Commerce, National Oceanic and Atmospheric Administration), p. 1.
4. Ibid., p. 6.
5. Ibid., p. 7.
6. James W. Beever, III, "The Effects of Fringe Mangrove Trimming for View in the South West Florida Aquatic Preserves," September 21, 1988, Report for Florida Department of Natural Resources.

CHAPTER THREE

✦

A BALANCED LOOK AT CONFUSING ISSUES

E nvironmental issues provoke emotional responses. Self-serving environmentalists have created crises where there were none. Unprincipled reporters have used rumors to panic citizens and elected officials. And others have muddied the water with special interests. Rather than improving humanity's environmental conscience, this abuse has more often just turned people against activities that create obvious impacts on the environment, from open-pit mining to commercial fishing.

We have often believed inaccurate information and sometimes have shamed ourselves with militancy. We have not been shown all the truth. This is why so much environmentalism has proven useless. Rather than identifying genuine problems and addressing them systematically, we have labored to eradicate symptoms of a disease that keeps spreading. Many people do not even know what their environment includes.

Our planet's ecosystem consists of three primary components: air, water, and soil. These components are so inextricably woven together that what we do with one affects the others. Pollutants discharged into the air reappear in the water and soil. Runoff carries pollutants from land into streams and lakes. This basic relationship defines how we manage our environment. Unless we consider all the impacts from our activities, we cannot comprehend the ubiquitous effects of the way we live.

One volume cannot address every piece of the environmental puzzle. Nor can one volume offer solutions for every problem. This chapter, however, looks at five issues that encompass our most common environmental problems: air pollution, water pollution, wetland destruction, solid waste disposal, and endangered species protection.

We cannot afford to wrap these issues in inflammatory rhetoric and political partisanship. If we look at the world with open eyes and clear heads, we will see measures we all can take to achieve a balanced environmental conscience toward the environment. We will not be pawns of those with selfish motives.

AIR POLLUTION

We humans pollute the air more universally than we do any other resource. Smog chokes large cities. Pollutants fall back to the earth in acid rain. Chloroflourocarbons may destroy ozone in the atmosphere and allow more ultraviolet rays to reach our skin. And air's habit of moving across national boundaries confounds an already complex problem.

Air pollution can be more than a nuisance. It can kill. In 1952, London experienced several days of meteorological stagnation. During this time, nearly 4,000 premature deaths were attributed to the extreme air pollution levels.[1]

Fortunately, air pollution problems, especially smog, received early attention. Air quality is still a critical concern. But many industrial nations seem to be less overwhelmed with air pollution today than they were twenty years ago. Many less developed countries, however, suffer extreme air pollution problems. Mexico City, for example, may choke beneath the worst smog level of any large city in the world.[2]

Mexico City does not suffer alone. Air pollution, like other types of pollution, relates to economics. Activities associated with population growth and industrialization generate air pollutants from fossil fuel combustion and industrial emissions. Minimizing these emissions takes money and makes economic development more costly. Poorer countries often must choose between development and pollution control. Many of these countries tolerate increased pollution levels in order to encourage development. But because air moves around, their pollution can become ours, and vice versa.

In addition to air's gadabout nature, other factors complicate measuring air pollutants, especially those related to health risks. Scientists know air pollution contributes to the incidence of lung cancer, emphysema, and bronchitis. But how much? Cigarette smoking complicates this question because smokers generally are more susceptible to those diseases than nonsmokers. And recent studies have shown inhaling another's cigarette smoke can be dangerous even to nonsmokers.

The human lung possesses an immense surface area, about ninety square meters (over 960 square feet).[3] This huge area permits rapid absorption of many substances from the air into the blood stream. Substances which are highly soluble in water may pass through the lungs so rapidly they are undetectable once inhalation ceases. Other substances may remain in the lung so long they cause inflammation, emphysema, malignancy, and other troubles. The ease with which we can breathe in foreign substances suggests inhalation as the most important route of entry of injurious substances into the body. Air pollution is a serious matter.

Since air pollutants interact with one another, assigning

responsibility for a certain problem may yield little more than an educated guess. For instance, sulfur dioxide interacts with particulates and reduces the ventilatory ability of the lungs.[4] Is sulfur dioxide or particulates the problem? We know this combination can stress our bodies enough to make the difference between life and death for someone with respiratory problems. But we hardly agree about which pollutant bears the greater blame.

We find it difficult even to define clean air. "Normal" dry air contains the following components (in order of concentration): nitrogen, oxygen, argon, carbon dioxide, neon, helium, methane, krypton, nitrous oxide, hydrogen, xenon, nitrogen dioxide, and ozone.[5]

We may consider any substance causing air to deviate from this norm to be a pollutant. But we rarely, if ever, find such air in nature anyway. This definition also does not consider the nature of the "pollutant" or its effect on humans or their environment. We might better define an air pollutant as a substance occurring in sufficient concentration to cause an unwanted effect.[6]

People do not produce all the pollution they breathe. Forest fires generate copious amounts of smoke. Pollen and dust can make people sick. And there are other natural sources of air pollution.

Regardless of its origin, air pollution occurs as either gasses or particulates. Ozone, hydrocarbons, carbon monoxide, and other gasses often occur in sufficient concentration to pollute the atmosphere. Nature generates many of these gasses. People, however, concentrate such gasses through waste disposal, industrial processes, fossil fuel combustion, and other means.

Particulates, too, are produced by nature and by people. These solid and liquid particles enter the air through the same vehicles that convey gaseous pollutants. Air pollution, then, mostly begins on land.

Today, industrialized societies judge air pollution primarily according to air quality standards. This point of view, though somewhat arbitrary, helps evaluate emissions in spite of the dispersing effects of air movement. We may thus distinguish between primary and secondary pollutants. Primary pollutants are emitted directly to the atmosphere while secondary pollutants are formed in the atmosphere.[7] Photochemical smog is secondary pollution.

Relying strictly on air quality standards, however, may blind us to serious consequences of air polluting emissions. Burning sulfur-containing coal affects rain, for instance. Rain naturally has a pH of about 5.6, slightly acidic because atmospheric carbon dioxide dissolves in raindrops to form carbonic acid.[8] But sulfur and nitrogen oxide emissions can cause acid rain. The more acidic the rain, the greater pH influence the rain has on soils, vegetation, and water bodies.

Acid rain has been blamed for killing fish and aquatic plants in 50% of the high mountain lakes in the Adirondacks of North America. Many of these lakes are reported to have reached pH levels so low as to replace normal aquatic life with acid-tolerant mats of algae.[9]

Weather events do not stay put, either. Storms traveling over the industrial areas of Great Britain and continental Europe have dumped particularly troublesome precipitation on Norway.[10] No longer can any of us afford to think we live on an island.

Worldwide today, we can emit into the atmosphere practically every form of pollutant imaginable. Even radioactive pollution threatens lives far removed from the source.

Following a massive explosion at the former Soviet Union's Chernobyl nuclear power station in 1986, everyone within a thirty kilometer (almost twenty mile) radius of Chernobyl was evacuated. Still, thirty people died. Farther away, radioactive rain showered reindeer herds and the lichens they eat in Norway and Sweden. Since then, monitoring has determined some herds to be too radioactive for human consumption.[11]

The health effects of air pollution show us the need for a balanced environmental conscience. The "boom times" perspective toward development was never conducive to a healthy human environment even though it boosted the human economy. Equally harmful has been environmental militancy.

Between these extremes lies a conscience open to new ideas yet recognizing no one can evaluate pollution with 100 per cent accuracy. Some scientists may think themselves almost infallible. Yet, current analytical techniques are not precise. "Pollution" depends many times on the effect a substance causes rather than on the presence of the substance itself. This makes measurements of environmental quality largely reasonable estimates rather than absolute values. Understanding this fact helps us see air pollution more clearly.

Often we waste money and energy trying to clean up emissions. We forget the least expensive and most effective pollution control occurs at the farthest point up the process line. The beginning of the process or alternatives to the process provide the most effective sites for pollution control.[12]

This truism applies to automobiles as well as to industries. For years, tetraethyl lead was added to gasoline as an antiknock compound. Lead particles were emitted with the smoke from automobiles' internal combustion engines. After cars became widely available, people grew to depend on this transportation so much that motor fuels with lead additives became the primary source of atmospheric lead. But lead is more than a pollutant. Lead interferes with red blood cell development making lead a serious potential health risk. The risk is even greater for urban dwellers and cigarette smokers.

Manufacturers have made great strides in controlling pollution from automobiles. We educated ourselves about pollution from our favorite mode of transportation and demanded solutions. Lead-free gasoline is one solution. Improved mileage efficiency is another. We can attribute these steps to an environmental conscience quite different from the conscience of the 1950s. With such a conscience, we can fight even smog, the scourge of large cities.

Photochemical smog forms in the atmosphere in a dynamic process involving sunlight plus nitrogen oxides and hydrocarbons from fuel combustion. Various compounds are produced, including aldehydes, organic acids, and epoxies. A potentially dangerous component of smog is ozone, a secondary pollutant.

The stratosphere also contains ozone. This is a different matter. Ozone in the stratosphere absorbs much of the sun's ultraviolet radiation and acts like a shield for life on earth. Stratospheric ozone is thought to be maintained primarily by solar radiation.

Ozone also occurs naturally in the lower atmosphere. Here lightning is the principal natural source of ozone.[13] Excess ozone seems to be caused by humans.

Minute amounts of ozone can cause pulmonary congestion, edema, and hemorrhage, and can reduce breathing capacity.[14] Ozone also can cause tissue collapse and other injuries in plants, including vegetable crops.

At one time, scientists thought reducing hydrocarbon emissions would reduce atmospheric ozone levels. This tactic failed. The answer to the ozone problem seems to lie in controlling *all* the primary pollutants involved in smog formation.[15] To do this, we consumers need to know how we affect the environment and how the environment affects us. We can then build on our success of reducing automobile lead emissions and reduce other primary air pollutants.

Perhaps we underestimated our ability to pollute the atmosphere. We need not underestimate our ability to clean it up.

WATER POLLUTION

1950's Japan: In the City of Minamata, 131 people display a variety of unusual symptoms of illness—progressive blindness, deafness, lack of coordination, and intellectual deterioration. Forty-six of these people die. Finally, it is learned a factory on Minamata Bay is discharging trace amounts of mercuric chloride in its effluent to the bay. This discharge alone can hardly cause the noted symptoms. But microorganisms in the bay's sediments convert the mercuric chloride to methylmercury, a highly toxic compound that can pass across the brain barrier in humans.[16]

The lesson? Even relatively harmless forms of mercury can enter the aquatic ecosystem and be converted to highly toxic forms. These toxic compounds can then pass along the food chain and be accumulated in fish and shellfish and, eventually, in people.

We humans greatly influence the world's water quality. Waste handling, sewage effluent, storm water runoff, and industrial discharges alter the quality of the water into which they mix. Human activities, especially urbanization, even affect precipitation. Yet, clean water remains a critical resource for our survival. Humanity's utter dependence on clean water accentuates problems associated with our water resources.

Water problems result from several factors. Water is mobile. It moves from place to place on and beneath the earth's surface and alters between gas, liquid, and solid states. Water's physical and chemical properties contribute to the many ways in which it can be polluted. Water supplies around the world vary from abundant to practically nonexistent. The unavailability of water in many regions may be the most critical worldwide environmental problem. Yet, water *quality* touches everyone.

Human activities in one place may affect water quality elsewhere. This situation imposes responsibility on each of us. We may not even live near the water. Yet, water pollution does not begin in the water. It begins on land. The way we manage land-based activities largely determines the water quality around us. Thus, we need to alter our environmental conscience to include other motives than just financial ones.

People make money by polluting water just as they do by polluting air. Not that anyone deliberately goes out to pollute. But preventing polluting discharges costs money.

For instance, properly designed and constructed storm water management systems reduce erosion, improve water quality, and attenuate downstream flooding. Yet, these ponds and waterways take up space that could be sold for building lots. Therefore, designers may minimize these facilities to maximize development. This attitude, though understandable, hinders long-range benefits from more responsible design and construction of storm water systems.

Other land-use practices affect our water, as well. Urbanization paves over land creating huge surfaces impervious to water penetration. These surfaces collect pollutants from automobile emissions, waste oils and greases, lawn and garden chemicals, and other sources. Then rain washes these pollutants into storm drains and ultimately to a receiving water body. The more pavement, the more runoff. The more runoff, the more pollution. And the rain has already picked up pollutants from atmospheric aerosols and particulates.

Agriculture, too, affects water. Irrigation consumes the greatest amount of freshwater in the world.[17] Modern irrigation technology has helped increase food production in the face of increased demand. But improper irrigation causes serious problems. Over-irrigation can waterlog soils and reduce yields. Salt water intrudes into overused groundwater aquifers near the coasts. Farms around the world are affected. By one estimate, as much as one-half of the irrigated soils in Syria suffer from excess salt content.[18]

Agricultural fields also contribute sediments and chemicals to storm water runoff, and these eventually wind up in receiving waters. Pesticides and nutrients percolate into the groundwater table, as well.

Human recreation also affects water resources. Motor boats leak oil and gasoline into the water and menace some species of wildlife, such as manatees. Boaters, beachgoers, and fishermen litter the water and its shores with trash. Plastics worsen the problem because they last so long.

Too often our attitude toward these problems has been, "it won't hurt me." To a degree we have been right. Nature's resilience has saved us from experiencing the full impact of our environmental blindness in our lifetime. But payday eventually arrives. Today we all suffer the burden of our predecessors' poor urban planning, irresponsible industrial development, and inadequate environmental enforcement. Unless we alter our own conscience, our progeny will likewise suffer from our mistakes.

Perhaps human conscience toward water resources has been harmed by our misconception of water's abundance. Water *is* the earth's most abundant resource. But the accessible and usable portion is very small. As the population of an area grows, the relative amount of available water dwindles. Just so much water exists in any area whether it is collected on the land's surface or pumped from the ground.

Shifting the blame for water problems will not work. No one is blameless, and our trust in "experts" and governments to solve water problems has proved faulty. We still need technical expertise and governmental oversight. But we all need to accept personal accountability in the ways we use our environment.

Since water pollution begins on land, many land-use practices need to change. This change *may* mean higher consumer costs in some cases.

For instance, chlorine manufacturers commonly use a mercury electrolytic cell. Some mercury is lost to the environment during this process, as much as 150-250 grams of mercury per kilogram of chlorine produced.[19] We know mercury accumulates in sediments and is concentrated as it moves up the food chain through predation. We also know mercury can be toxic to humans. Yet, a known substitute for the mercury cell in chlorine manufacture yields an inferior product at a higher cost.[20] Not only does this solution raise the price of chlorine, but it affects the prices of related products, too.

We should not automatically expect a change to raise prices, however. Modern tillage and crop rotation have increased crop yields and reduced soil losses, thus helping to keep food costs down. Sometimes, though, we can expect a cost to be incurred relative to land-use decisions we make, good or bad. The question is when we will pay.

During good economic times, we may restrict economic growth to conserve essential resources for our long-term benefit and the benefit of those who follow us. But a deteriorating economy threatens even the most essential elements of the environment. So, we need to get in the conservation habit today.

To formulate a conservation perspective toward water resources, we need to understand at least a little about the hydrologic cycle. Water recycles between the oceans, the atmosphere, and the land. Scientists believe that most of this water has existed since early in the earth's history.[21] That means we drink water that is centuries old. However, water's residence time in any single location varies considerably.

Water stays in the atmosphere an average of only ten days but may reside deep in the ground for 10,000 years.[22] Precipitation continuously moves to the earth's surface what humans put into the earth's atmosphere. And groundwater has ample time to accumulate chemicals penetrating the earth's surface.

Even though federal and state governments enact water quality standards based on designated uses, water quality can be a nebulous goal. Standards often reflect minimum conditions necessary for the intended use, rarely optimum. If we act just to meet those standards, we may force the water to be minimally suited to its expected use. Then, when more trouble comes along, such as a hurricane, fuel spill, or forest fire, little buffering capacity exists within the resource itself. Fish may die, beaches may close, and recovery may take a long time.

This prospect can hit us where we live. A popular tourist city in Florida gets its drinking water from a small reservoir in a rapidly developing watershed. The reservoir lies outside the city limits, and city officials have little impact on development within that watershed. Algae blooms, dissolved oxygen problems, and fish kill complaints commonly occur in the reservoir and its tributaries. Yet, state water quality standards normally are met.

A conscience that balances each element of the ecosystem and the various benefits of the resource would have helped prevent some of the problems with this city's water supply. So, what is the real problem there? Remember, the love of money is the root of all kinds of evil.

People are not a curse. But human activities greatly speed up processes that naturally take a long long time. For instance, lakes form, age, and eventually disappear. Natural succession takes centuries, but erosion, urban runoff, and industrial effluent can reduce these centuries to decades. In time, some of the values that attracted people and industry to the area may disappear, too. By that time, though, most of the people responsible for the state of affairs may be gone. That is why we need a far-sighted perspective if we are to manage our environment in a responsible manner. Solutions require cooperation.

Partisanship will never solve environmental problems. Neither will panic. Acid rain provides a good case in point. Environmentalists have implicated acid rain in a number of adverse effects on soils, water bodies, and agricultural crops. But acid rain is no cause to panic. Rain is naturally slightly acidic. However, parts of Canada and the United States are experiencing increasingly acidic precipitation. Studies indicate air pollution by each country causes acid rain in the other. And these countries are not unique. The pronounced acid rain phenomena in Scandinavia has been attributed to sulfur oxide pollution from Western Europe.[23]

No single solution exists to a problem crossing international boundaries. But a cooperative attitude generates motivation and expenditures needed to resolve this problem to the benefit of all participants. Each country's "me first" attitude must die, however.

In earlier years, people looked for an industrial culprit whenever they noticed a problem with their water. We no longer enjoy that luxury. Industries sometimes do pollute water, but we all contribute to the problem.

Everyone produces waste water (sewage). The more concentrated the population, the more waste water produced. We have to put the effluent and sludge somewhere. They do not just go away. This fact of life means we must clean up the waste water and release it to the environment.

However, waste water treatment costs a bundle. The more extensively waste water is treated, the more expensive treatment becomes. Consequently, local governments may resist building advanced waste water treatment plants because sewage rates might go up. Public utilities may truthfully point to an effluent that is no direct threat to human health due to adequate disinfection. But pathogens are not the only "pollutant" in waste water. Waste water effluent contains nutrients. It can stimulate vast algae blooms in fresh and salt water. Many times, these blooms look awful and stink. They only rarely create a true health risk, but they often diminish the environmental quality of the water body by increasing turbidity, depleting dissolved oxygen, and shading out benthic organisms. In turn, oxygen depletion sometimes kills fish. Yet, most people do not understand the problems with handling waste water. They simply flush it away.

Domestic waste water forms but one piece of the giant water quality puzzle. Runoff from paved surfaces, urban landscapes, and agricultural fields pummels waterways with metals, nutrients, pesticides, and sediments. Industrial emissions still pollute water as well as air. Lax regulatory programs and outdated water quality standards encourage degradation of the very waters they are expected to protect. And most people do not understand the scope of the water crisis, though they contribute to it.

We have grown up discharging liquid wastes into the water. Oceans, in particular, have been thought of as great sinks that solve pollution through dilution. But oceans are fragile. Few of us comprehend the complexity of ocean water. Many scientists believe the chemical makeup of ocean water has changed very little over untold centuries. This consistency has helped many marine organisms develop into highly specialized creatures tolerant of only minor environmental change. Degradation of ocean water quality easily stresses this fragile ecosystem.

In spite of the oceans' fragile nature, countries the world over still pipe sewage and industrial wastes into the oceans. Cruise ships and navies still dump garbage overboard on the high seas and sometimes just before entering port. Cities still channel runoff directly into bays and estuaries. And boaters and fishermen still throw their trash into the water.

We even pollute groundwater. Some of the water from precipitation seeps into the ground. Eventually, a portion of this water reaches the zone of saturation. We can tap the water in this zone. In water's path to the zone of saturation, it picks up nutrients, minerals, pesticides, gasoline, bacteria, and just about everything else we disperse in and on the land. We should never think of soil as a water filter.

We all have heard horror stories of chemicals detected in or near drinking water supplies. It really does happen. Yet, groundwater contamination is not fully understood even by the experts. Problems exist in identifying, prioritizing, and remediating such contamination. Even in a technologically advanced society, groundwater contamination can cause a permanent loss of a water resource. Thus, an effective groundwater management strategy must prevent degradation.

Most of the United States still enjoys good quality groundwater, though its availability varies. But we should not feel too safe. Not all groundwater can be tapped. Nor should we withdraw too much water from one aquifer and deplete the resource. This foolishness creates a host of problems of its own.

We face the same problem in conserving groundwater as we face in conserving other natural resources: we produce more waste than the environment can absorb on its own. We dispose of garbage, hazardous wastes, sewage effluent, and other wastes. Each of these agents can seriously pollute some portion of the environment.

In recent years, we have begun injecting liquid waste into deep wells. Like landfills, these systems must be carefully designed and operated for a long, long time to be safe. But waste disposal forms only one source of potential groundwater contamination.

Humans bathe the land in chemicals, and a certain amount goes into the ground. We also over-pump some aquifers. Near the coasts, over-pumping allows salt water to intrude into the groundwater, sometimes making wells unusable.

Fuel spills can contaminate groundwater. By design, septic systems inject large volumes of waste water into the groundwater. Even road de-icing can contaminate groundwater.

So far, though, identifying the extent of groundwater contamination has proven difficult. In addition, our limited understanding of groundwater systems leaves many questions unanswered. As is often the case, what we do not know poses one of the greatest problems in managing a resource.

We can only guess at the long-term consequences of our race to pollute. The Bible teaches accountability even when the end result of our actions outlives us. Practicing accountability will enable us to conserve resources, including precious water, for our use and the use of future generations.

WETLAND DESTRUCTION

Wetlands exist because the water level is at, near, or above ground level much of the year. Wetlands take many forms: bogs, marshes, prairies, hardwood swamps, cypress strands, mangrove forests, riverine floodplains, and more.

People alter or destroy wetlands to make money. They want more farmland to grow more crops, more pastureland to raise more cattle. They want a better lot to sell for a house, a convenient site for a shopping mall. They want more and wider roads to move more people and goods faster. A wetland may just fill the bill at a bargain price. Or so it seems.

None of these enterprises is necessarily wrong. But an irretrievable loss occurs when these activities are carried out in a wetland.

Historically, wetlands were considered wasteland. The only way people thought they could produce a valuable property from a wetland was to drain it or fill it. Coastal estuaries became subdivisions. Inland swamps became industrial plants. River floodplains became farmland. According to the United States Congress, the United States has lost 30-50% of its wetlands to mining, forestry, agriculture, and urban development over the past 200 years or so.[24]

We have produced additional wetland impacts by our activities on nearby uplands. Any activity that removes vegetation from the soil's surface accelerates erosion which transports sediment to wetlands and surface waters. In addition, water diverted from wetlands can intensify drought conditions. Not only have we ignored the intrinsic values of wetlands, we have not always understood those values.

Healthy wetlands benefit everyone. Wetlands collect water from runoff and improve its quality by allowing solids to settle out and by trapping nutrients and other pollutants. When wetlands are reduced through draining or filling, travel time of surface water flow is also reduced. This, in turn, reduces the amount of filtering provided by the wetland.

By slowing runoff velocity, wetlands attenuate flood flows. The type and size of the wetlands determine how much flooding is affected, but wetlands can reduce flood peaks. When wetlands are developed, expensive measures must be taken to control flood peaks. These dams, dikes, canals, and channels function at the expense of water quality and habitat values. And someone has to pay for them.

Wetlands also provide critical habitat for fish and wildlife. A great variety of commercial and sport fish and shellfish need wetlands for spawning and nursery areas. Waterfowl use wetlands as resting, nesting, and wintering grounds. Hunters are keenly aware the future of waterfowl hunting depends on conserving wetlands. Conserving wetlands is also crucial to many endangered species. Twenty percent of all the plants and animals listed as threatened or endangered in the United States require wetland habitat.[25]

Obviously, reducing wetland size reduces habitat. But altering a wetland also can lower the value of the wetland that remains. Quality of habitat can be as important as quantity when threatened or endangered species are involved. The more restrictive a species' habitat requirements, the more that species relies on specific conditions. Thus, stress from development can lower the value of wetlands for many creatures, including humans.

All these values of wetlands share a common problem: the difficulty of assigning a monetary value to them. An exception may exist in some forested wetlands. Timber grown on these sites may generate a substantial income that would be lost if the wetland was destroyed. Still, most wetland values rarely stand before the short-term economic advantage of development.

Perhaps this perception is changing. Americans may be waking up to the fact each of our water resources deserves protection. One day we may have to drink it. It helps, too, that American society is affluent enough that we have made wetland protection a political issue.

Federal and state laws protect wetlands to some degree. Generally, though, Americans have agreed development in wetlands should continue as long as negative impacts are minimized, or if that is impractical, mitigated. Effective mitigation, however, has often proved an elusive dream.

Developers create wetlands from uplands in hopes of mitigating the loss of natural wetlands. By doing so, they venture into the unknown. None of us understand all the factors that work together to form a successfully functioning wetland. Permissive mitigation policies risk actual wetland loss in spite of legal compliance.

Overemphasizing mitigation also blinds us to the fact we are eradicating one ecosystem to create an entirely different ecosystem. We may wake up one day and realize some types of uplands have been just as important as wetlands. Only by then we may have lost many of these uplands to artificial wetlands.

Careful planning and a willingness to settle for less profit can reduce the need for mitigation by reducing wetland loss during development. Emphasis will then shift from wetland mitigation to wetland protection. A conservation ethic practically demands protection for such a productive ecosystem.

Wetlands have developed over time into ecosystems of particular plant and animal communities tied to specific hydroperiods. These water regimes and soil characteristics do not always lend themselves to duplication by humans. A functioning wetland filled for a housing development may be lost forever even though its loss was legally mitigated by the creation of a wetland elsewhere. But we have not learned our lesson. Technology cannot replace nature.

Mitigation merely displays human arrogance. It presumes we know what we are destroying and know how to create something just as good. We do not. Overstressing mitigation achieves only one result, a corrupt permitting system. It can come back to haunt us, too.

The Kentucky Transportation Cabinet apparently had no idea it would create a salt water wetland in northeastern Kentucky while mitigating for a natural wetland lost to highway construction. But in 1987, they did just that.

A slough connected to Salt Lick Creek was filled and a new wetland created and connected to the creek. Workers had already conducted extensive faunal and floral surveys. Aquatic fauna were moved from the old site to the new one, and the organic seed-bearing muck from the old site was distributed around the border of the created wetland. After all this work, a local resident noted the new wetland had been constructed on the site of an old "salt mine"—a salt-laden spring where Kentucky pioneers had collected salt by boiling away the water.[26]

Chloride levels in the created wetland shot to over 35 times the chloride level of the wetland that had been filled. Solutions to the problem were deemed expensive or nonexistent. Nature, however, rescued the state. The translocated fauna survived, and winter rains diluted the salt. Eventually, siltation or some other process reduced the input of salt-laden groundwater. Chloride levels then stabilized just a little above the pre-construction level of the filled wetland.[27]

While a salt marsh fits right in with coastal Georgia, a salt marsh is foreign to Kentucky. And, in this case, no engineer or biologist could correct the mistake. Nature's resilience rescued humanity from its own pride and ignorance. Still, we find it difficult to acknowledge the far-reaching values of natural wetlands. We also hate to admit our ignorance at successfully constructing and maintaining such functional ecosystems. And "functional" is a key word.

Studies show net primary productivity of freshwater wetlands equals that of tropical rain forests. Salt marshes exceed that. However, wetland destruction has not generated the same worldwide attention as rain forest destruction. The tragedy of the rain forests dramatizes the fallacy of applying a narrow economic view alone to environmental issues.

We should not deny economics, however. Some environmentalists have cried "wetland" to block development even when no reasonable excuse existed. Much like the boy who cried "wolf," this deception has numbed many people to all wetland protection issues. In the long run, we all can lose—lose wetlands, lose money, lose credibility.

To reduce this economy/environment conflict, we need to weigh the economic impacts of environmental decisions. Each environmental issue includes an economic issue. Approaching environmental decisions with no economic perspective eliminates the principal agent in resolving the conflict: people. This practice may preserve a wetland (or may not), but it will not resolve the conflict.

Looking at wetlands with an economic eye alone also will never solve the problem. We need equally to approach wetland issues from an environmental perspective. Such an attitude prompts us to examine alternatives and their effects on all elements of the ecosystem. This approach may also increase the cost of some, but not all, development. The tradeoff, however, will be to increase the natural benefits we derive from the environment. Without this approach, environmental management consists of reacting to crises resulting from our lack of foresight. A healthier approach heads off many crises because we look at more factors from more angles.

A greater challenge? Certainly. But an ecosystem viewpoint improves environmental management.

Habitat value, for instance, suffers when compared to the financial value of development. We figure habitat loss can be mitigated if that habitat warrants it. Thus, permitting requirements supersede ecological principles.

By considering all the values of a wetland, we may learn

how to derive greater benefit through conservation of the resource. But that can be a real challenge. Short-term profit may be reduced in favor of long-term benefits. For this tactic to succeed, our environmental conscience must embrace a viewpoint other than a monetary one alone.

We cannot just tack on the price of some use, such as a hunting lease, as an economic value. That strategy will never defeat a genuine business perspective. We would still be fighting a financial battle, and a hunting lease will almost never win that war. But other values can compete with development.

Probably everyone knows good water quality is crucial to healthy people and to a healthy environment. Wetlands improve water quality cheaply. This fact is so well documented some environmental managers propose using freshwater wetlands to treat effluent from sewage treatment plants. The long-range effects of this use require more study. We know for sure, though, wetlands improve the quality of the water entering the wetlands in several ways.

Water movement slows down. Thus, the erosion potential lessens, and suspended solids settle out. Further mechanical filtering occurs as water percolates through the soil. In addition, plants extract nutrients from the water. Nutrient removal is very important to downstream receiving waters since excess nutrients stimulate algae growth. Algae blooms can kill fish, cause unpleasant odors, and increase the treatment cost of drinking water.

Nutrients dissolved in the water require treatment for human consumption. The more treatment accomplished prior to water entering a treatment plant, the less treatment will be necessary in the plant. Thus, the consumer saves money. We will not save anything, however, if we do not recognize wetlands as *Very Important Properties.*

Freshwater sources and wetlands comprise perhaps the world's greatest natural resource after people. Stress from development threatens these waters and threatens their value for humans, making wetland conservation an important matter for everyone. We need functioning wetlands.

SOLID WASTE

In 1657 in New Amsterdam (later, New York), residents dumped ashes, dead animals, and other waste into the streets, much to the chagrin of city officials. Human waste contaminated shallow wells and caused catastrophic cholera epidemics.[28] Over 300 years later, Americans are still trying to get rid of a growing mountain of garbage. This solid waste invokes visions of overstuffed landfills, highways strewn with litter, and dead sea birds entangled in plastic. But modern waste problems, though often exaggerated, are more complex than those visions.

Various reports over the past decade or so have suggested Americans throw away garbage in amounts ranging from three to eight pounds per person per day. However, studies performed by the Garbage Project at the University of Arizona indicate even three pounds may be too high a figure in much of the country.[29]

Many environmentalists have cried loud and long about pollution from solid waste production and disposal. This clamor has attracted much public and political attention, and for this we can be grateful. Like it or not, we must learn how to live with our garbage. Too often, though, we have emphasized the less important waste problems because these have been the most visible. Doing this makes us think we are more wasteful than we actually are and more wasteful than we were in the past.

Every day we see fast food packages littering the roadside and lament America's love affair with plastic. We see junk yards and decry the business strategy that designs rapid obsolescence into manufactured goods. But we have not grown in wastefulness. Our waste has changed.

Modern technology, such as gas and electric heating, has eliminated the 1,200 pounds per person per year of coal ash once dumped throughout the United States. Modern transportation has replaced the hundreds of thousands of dead horses that once plagued American cities. Modern food packaging prevents much food waste. But we do not see the wastes we no longer produce.

Consequently, we appear to produce more garbage than we actually do produce. The Garbage Project suggests America's per capita production of household garbage is relatively stable.[30]

This stability gives us breathing room. Instead of blinding ourselves with exaggerated volume claims, we can address solid waste problems from a more rational point of view.

When most of us think about solid waste, we think about municipal solid waste (MSW), the solid and semisolid wastes from a community. Refuse constitutes that part of MSW from households. Sludges, nuclear wastes, and hazardous materials also make up solid waste. But even if we consider only MSW, we still can see solid waste requires proper management to avoid environmental harm and human health risks.

For many years, Americans have used landfills for "sanitary" disposal of MSW. But environmental and health risks exist at any landfill. Liquid leachates can enter groundwater aquifers and contaminate potable water wells, as happened with the Town of Islip's landfill on Long Island.[31] Rats and flies can breed prolifically around these sites and spread disease. We know of some 22 human diseases associated with solid wastes.[32] Animals, such as sea gulls, feeding in landfills sometimes die from what they eat. We find it increasingly difficult to dispose of solid wastes in a way that removes all impact to society.

We should not wonder. Problems with solid waste arise from its source, collection, and disposal. To solve the solid waste puzzle, we must work with all the pieces.

Popular environmental thinking usually has focused on waste generation. As one solution, many people favor recycling. Recycling does sidetrack from the waste stream a certain amount of a limited number of materials. But sidetracking is temporary. Eventually, even recycled materials are thrown away. Mandatory recycling may even counter more effective commercial recycling efforts.

Source reduction also has been pushed as the answer to the problems of some types of solid waste, most notably plastic. In theory, a waste problem is eliminated before it occurs. But an undue emphasis on source reduction distracts our attention from more beneficial measures.

Consumer industries already make their products as compact as practicable for their use. It saves them money. A typical plastic soda bottle weighed sixty grams in 1970 but weighs only 48 grams today.[33] And consumer products still must be packaged in something.

Too much emphasis on source reduction also may focus too much attention on plastics. Plastic six-pack rings can trap birds and other wildlife. Plastic bags and wrap can kill sea turtles that mistake them for jellyfish and eat them. Plastic littering the highways offends our aesthetic senses. But source reduction will not eradicate the slovenly attitude that prompts people to litter their environment. Plastics reduction also will not add much life to a landfill.

Researchers with the Garbage Project examined 16,000 pounds of garbage from seven sanitary landfills in the United States. They found plastics of all kinds taking up only twelve per cent of the landfills' volume and less than five per cent of the landfills' contents by weight. A phobia about plastics has diverted people's attention from the real culprit gorging landfills. Paper comprises forty to fifty% of everything thrown away.[34]

Only recently have many people recognized the more serious problems with landfills. Many landfills still in use were designed for simpler times and sited for less hazardous materials than what we dump in them today. Some landfills still are nothing more than open dumps.

Following World War II, landfills were often located in wetlands, one of the worst possible places for solid waste disposal. Households now use oils, pesticides, and cleaning fluids that in past decades were largely confined to industry. The risk of chemical leachates contaminating water supplies gives us cause to worry.

Usually, we attack solid waste problems through peripheral issues, such as fast food packaging. The real obstacles, though, are money and psychology. Even public health concerns rarely decide how and where we dispose of solid waste. We want the cheapest means to get garbage out of our way.

Most solid waste is collected by truck. Transporting garbage by truck is expensive, and locating a landfill close to its "supply" may keep garbage collection fees from rising. We all like that idea. On the other hand, nobody wants to live next to a dump. The fact remains, however, the right kind of land for landfills is available. We may have to pay more to reach it, and someone will probably have to live nearby. But our environmental conscience will determine whether we are part of the problem or part of the solution. Sooner or later, each of us will face such a choice.

Our conscience relative to solid waste needs to emanate from the present and look to the future. Open dumps provide no sanitary waste disposal today, nor will they tomorrow. We need to give high priority to design and location of landfills if we are to dispose of our wastes in a manner conscious of our environment and our health.

Economics need not be ignored when planning for landfills. But overemphasizing transportation costs can cloud our judgment toward other vastly more important matters, such as length of time of usefulness, impacts to air and water quality, and future uses of the site. The economics of these factors may easily dwarf transportation costs.

Landfills usually provide the least-cost method of acceptable solid waste disposal.[35] Still, landfills address only this one facet of the solid waste puzzle. We must address all three facets in order to solve the problem over the long run.

Many people now see solid waste as a pollution problem on a par with air and water pollution. Too often, though, they think a single strategy, such as recycling or biodegradable packaging, will win the war. But no single solution exists.

Our conscience determines our lifestyles. Our lifestyles determine the type and amount of waste we produce. Increasingly, waste disposal risks harm to the environment and to human health in excess of the cost of transporting that waste.

These risks do not have to bear fruit. Our planet is resilient, and humans are adaptable. We can balance resource use, packaging concepts, manufacturing goals, and disposal economics to create a climate conducive to effective resource management. Our conscience will decide how well we live at peace with our environment.

ENDANGERED SPECIES

Few environmental topics generate as much heat as the subject of endangered species. Some people want to preserve every plant and animal on earth at whatever cost. Others view many species as roadblocks to their dreams. Perhaps one persuasion balances the other, but a danger lurks with both.

The first group may consider humans intruders. These vile creatures, they suppose, invade existing ecosystems bent on destruction. The second group may consider themselves masters of all they survey, seated above the ecological systems operating in the world. One group sees humanity as the enemy. The other group sees humanity as a god.

Neither persuasion bears good fruit. The resulting conflict can be compared to war where even the winner loses a lot. Somewhere between these extremes lies an ethic that recognizes the benefits of a healthy environment as well as the needs of the human element of that environment. People are not always to blame.

Many of us accept the biblical account of Creation and the great Flood. In this context, dinosaurs may have been destroyed in the cataclysmic flood except for those few preserved in the ark. But the earth's environment, including its climate, had changed so drastically following the cataclysm that the dinosaurs could not adapt, and they died out.

Many people reject that theory of origins in favor of an evolutionary theory. In this context, dinosaurs gradually evolved into higher forms of life, thus changing with their world. Again, those species that could not adapt died out.

Regardless of which view one accepts (or another one), the fact remains dinosaurs are extinct. The environmental changes that led to their extinction were not designed and produced by humans. In modern history, however, extinction has often come through human aggression toward a species or through the inability of the species to adapt to alterations of its environment.

The Carolina parakeet once ranged from Florida to New York and Illinois. This bird was common, and its feathers were valued by milliners for hat trimming. Carolina parakeets were hunted at a time when game laws were in their infancy, and people rarely noticed potential environmental problems. Today these birds are extinct, the last flock reported in the Florida Everglades in 1904.

Humans survived the passing of both dinosaurs and Carolina parakeets. We may even be glad we do not have to share our home with a formidable predator like Tyrannosaurus. But a dull conscience toward such matters devalues any part of the environment that cannot easily survive. This attitude burdens certain elements of the environment far beyond what they were designed to endure. And we humans can be the losers, too.

The human element of the environment reacts to the same stimuli that affect lesser creatures. Some of us do not adapt well, either. Are we, then, to pass away because an inherent defect prevents us from overcoming the pressure and aggression we experience? Certain of our brethren would answer, "Yes. If you can't cope, you should get out of the way."

Consider the outcome of this attitude. Cities around the world are littered with drug addicts who could not face the rejection of family and peers. Inner city neighborhoods reel under the onslaught of crime so violent and degrading human beings are reduced to being property or plunder. Prisons are so crowded dangerous criminals are released early to make room for new arrivals. Babies are killed by the thousands because mothers find it inconvenient to carry them to full term. And people willingly pollute their own environment to make money. Somehow we forget the quality of human life can be as endangered as any plant or animal.

We know nature abhors a vacuum. When one species disappears, another species or ecological system fills the void. But this tendency does not mean the new order will work the same as the old or that no net loss will occur.

People in Europe and the United States may feel little loss if the Javan rhino disappears. People in Indonesia, however, probably will feel a great loss. In 1921, Indonesians established the Ujung Kulon National Park to protect three rare species of mammals: the banteng bull, the Javan tiger, and the Javan rhino.[36] The Javan tiger did not survive.

Managing endangered species requires a strategy different from other resource conservation. Many, perhaps most, of these plants and animals need such specialized habitat they cannot adapt well to a rapidly changing world. Any extra stress, such as poaching or critical habitat loss, may push any of these species past the point of no return. For every creature, a population limit exists below which species perpetuation is doubtful. Some rare species probably never were abundant.

Habitat preservation and physical protection may offer some species their only hope for survival in the wild. The caterpillar of the regal fritillary butterfly, for instance, feeds only on the Kansas prairie violet.[37] This butterfly's survival depends on the violet's preservation in sufficient numbers to support a stable butterfly population. Probably nothing less will do.

Some people, though, reject the whole notion of endangered species. To them, this issue is no more than an exaggerated cry of panic. But consider a few ecological facts.

1. Many animals establish territories to support themselves and their young. These species do not do well when crowded.
2. Only so much food and cover exist in a given area. This supply determines the carrying capacity of that area.
3. A minimum population level must remain to assure enough reproduction to offset mortality and maintain a stable population.

4. The earth is finite.

Bald eagles, indigo snakes, California condors, and West Indian manatees have filled important ecological niches for generations. Their passing would leave voids in the hearts and minds of many people even though nature would adapt to their loss. This is the tragedy of endangered species. Because of a deficient environmental conscience, we risk losing a resource that cannot be replaced—ever.

We see this concept better with animals, but plants can face the same danger. Wild ginseng grew commonly throughout the temperate region of eastern North America as far west as the Mississippi River. Commercial harvest began in the early 1700s. The Chinese thought ginseng was a powerful therapeutic herb and bought large volumes of North American as well as Asiatic ginseng. Within a century, the North American plant faced near extinction. Today ginseng is scarce. We now know of no medical value of ginseng, yet it was harvested to the brink of extinction by people who sought it to sell as medicine.[38]

Even back then, economics shaped human conscience. Not that conscience alone will ever prevent extinction of anything. Conscience must stimulate action to bring results. But our action needs to follow a route plotted from an educated and sensitive conscience. This maxim holds true for any need but seems especially solid when applied to environmental issues. Our conscience determines how we live.

Can we really make a difference just because we want to? Our conscience, fine tuned through knowledge and determination, generates thoughtful action. Then, what appears impossible may become possible.

Sixty years ago, trumpeter swans clung to a shaky existence. The world's largest swan, they had been hunted intensively for their quills, feathers, and down. By the 1930's, less than 100 of these pure white birds with wings spanning eight feet were thought to exist in all North America.[39] This estimate is now thought to have been a little low. Still, this great bird numbered so few it had become a symbol of America's vanishing wilderness.

In 1935, the Red Rock Lakes National Wildlife Refuge was established in the Rocky Mountains of the United States to protect these swans. Other measures were taken in Alaska and Canada to protect remote nesting ranges and to reduce poaching. By the mid-1980s, trumpeter swans numbered about 10,000.[40]

Trumpeter swans signal an environmental success story. People had cared, learned about the swans, and taken appropriate measures to protect them. But these birds have survived in relatively remote regions of North America. Heavily populated regions also contain endangered species, but crowding encourages people to care less about a lot of things. Shaping a sensitive environmental conscience may require extra effort in an urban situation. But that is where it may be the most important.

Probably the greatest threat any endangered species faces is human greed and selfishness. A few years ago, American alligators were losing a two-front war of poaching and habitat loss. We considered wetlands valuable only to fill for development. We valued alligators merely for their skins. Since then, we have learned the economic as well as environmental value of wetlands, and we have recognized the ecological value of alligators. Wetland protection and tighter controls on harassment have created an environment where alligators now thrive.

But success did not come easily. Alligator hides, even illegal ones, bring a high enough price to tempt unscrupulous hunters. Many people still fight any restrictions on maximizing the profit from their land regardless of the long-term cost to anyone else. Ecological functions still bow to a quick dollar.

Our conscience dictates our love of money and justifies all manner of lamentable behavior. The end can justify the means, we tell ourselves. But we can train our conscience to be thoughtful, considerate of others, and content with a little less.

If we can win a few battles with our conscience, we can win the war. Victory requires we reorder our priorities and realize we do not exist for ourselves alone. Victory also requires timeliness. We need to learn the truth and then act on it.

NOTES TO CHAPTER THREE

1. Merril Eisenbud, *Environment, Technology, and Health— Human Ecology in Historical Perspective* (New York: New York University Press, 1978), p. 269.
2. Julie Sullivan, ed., *The American Environment* (New York: H.W. Wilson Co., 1984), p. 21.
3. W.M. Gafafer, ed., *Occupational Diseases—A Guide To Their Recognition*, U.S. Dept. of Health, Education, and Welfare, Public Health Service Publication No. 1097 (Washington: U.S. Government Printing Office, 1966), p. 8.
4. Eisenbud, *Environment, Technology, and Health*, p. 273.
5. P. Aarne Vesilind and J. Jeffrey Peirce, *Environmental Pollution and Control*, 2nd ed. (Ann Arbor, Michigan: Ann Arbor Science Publishers, 1983), p. 245.
6. Ibid., p. 243.
7. Ibid., p. 250.
8. Edward A. Fernald and Donald J. Patton, eds., *Water Resources Atlas of Florida* (Tallahassee: Florida State University, 1984), p. 84.
9. Vesilind and Peirce, *Environmental Pollution*, p. 248.
10. Ibid.
11. Charles E, Cobb, Jr., "Living With Radiation," *National Geographic* 175, No. 4 (April 1989): 423.
12. Vesilind and Peirce, *Environmental Pollution*, p. 313.
13. Ibid., p. 259.
14. Ibid., p. 258.
15. Ibid., p. 251.
16. Eisenbud, *Environment, Technology, and Health*, p. 258.
17. Fernald and Patton, *Water Resources Atlas*, p. 3.
18. Ibid.
19. Sachinath Mitra, *Mercury in the Ecosystem: Its Dispersion and Pollution Today* (Switzerland: Trans Tech Publications, 1986), p. 53.
20. Ibid., p. x.

21. Fernald and Patton, *Water Resources Atlas*, p. 10.
22. Ibid.
23. Eisenbud, *Environment, Technology, and Health*, p. 282.
24. Fernald and Patton, *Water Resources Atlas*, p. 95.
25. Ibid., p. 92.
26. Hal D. Bryan, "A Saltwater Wetland in Northeastern Kentucky," in *Proceedings of the 15th Annual Conference on Wetlands Restoration and Creation*, ed. Frederick J. Webb (Tampa, Florida: Hillsborough Community College, 1988), pp. 27-28.
27. Ibid., p. 28.
28. Eisenbud, *Environment, Technology, and Health*, p. 44.
29. William L. Rathje, "Rubbish," *The Atlantic* 264, No. 6 (Dec. 1989): 101.
30. Ibid.
31. Vesilind and Peirce, *Environmental Pollution*, p. 170.
32. Ibid., p. 153.
33. Rathje, "Rubbish," p. l08.
34. Ibid., p. 102.
35. Vesilind and Peirce, *Environmental Pollution*, p. 174.
36. Dieter and Mary Plage, "Return of Java's Wildlife," *National Geographic* 167, No. 6 (June 1985): 758.
37. Robert Murphy, *Wild Sanctuaries: Our National Wildlife Refuges—A Heritage Recorded* (New York: E.P. Dutton and Co., 1968), p. 35.
38. Alyson Hart Knap, *Wild Harvest: An Outdoorsman's Guide to Edible Wild Plants in North America* (Toronto: Pagurian Press Limited, 1975), pp. xi-xiii.
39. Charles A. Bergman, "The Triumphant Trumpeter," *National Geographic* 168, No. 4 (Oct. 1985): 547.
40. Ibid.

CHAPTER FOUR

✦

ENVIRONMENTALISM
IS NOT
THE ANSWER

Ecological principles rule the human environment. No amount of planning, engineering, or spending elevates our activities above the laws of nature. Consequently, the long range success of our endeavors hinges on how well we incorporate ecology into our activity.

The widespread realization of this interrelationship birthed the modern environmental movement in the 1960's. Many of us became suddenly concerned about pollution, conservation, and overpopulation. Environmentalism became an end in itself.

Environmental matters had attracted attention much earlier. T. R. Malthus urged population control as early as 1798.[1] The United States government, recognizing health and safety risks from coal mining, authorized federal standards for coal mines in 1890.[2] Los Angeles smog attracted public attention in the 1940's.[3] But widespread popular concern for the environment waited until the latter Twentieth Century to develop into a "cause." Then environmentalism caught on with a vengeance. Much good has come out of the modern environmental movement. In 1966, Congress passed the Clean Water Restoration Act. This act authorized federal funds to help local governments build sewers and waste-treatment plants.[4] The Environmental Protection Agency was created in 1970 to consolidate environmental functions scattered among many federal programs.[5] (EPA probably has grown too large, now). A number of citizens' organizations, such as the Audubon Society, have expanded their interest from wildlife to include other environmental concerns.

But during this great surge of excitement, a subtle flaw crept into our conscience. Environmentalists recognized their role as global ecologists. Too often, though, they failed to see that human environment extends beyond the narrow scope of clean air and water. This environment includes jobs, housing, urban sprawl, automobile traffic, and more.

For years, environmentalists have stressed reducing automobile tail pipe emissions. But automobiles affect human health, safety, and environment in far more serious ways, such as congestion, traffic accidents, and gargantuan sheets of pavement.

Environmentalists have seldom taken such a close look at urban and highway planning. And defective planning has exacerbated the negative effects of automobiles. This is one reason environmentalism by itself will never shape a truly effective environmental conscience. Environmentalism too easily misplaces its emphasis.

Environmentalism also divides populations rather than bringing them together. Exceptions exist, but environmentalism appeals more to people nearer the top of the social ladder. An extensive polling showed environmental concern picked up strongly at income levels between $30,000 and $70,000.[6] This data might help explain why minorities have been largely excluded from the environmental movement.

Some environmental policies help condemn the poorer underclass to permanent poverty. "No-growth" policies do not remedy the environmental problems of inner city slums and Third World destitution. Thus, our motives and emphases need to differ from those of many environmentalists if humanity is to achieve genuine environmental progress. We need to pull together.

Environmentalists have often criticized the effects of human behavior on the "natural" elements of the environment. Yet, they have rarely addressed the sociobehavioral environment surrounding us. This environment influences our health in profound ways.

Cigarette smoking generates the most noxious pollutant to which most Americans are exposed.[7] Even nonsmokers must inhale some tobacco smoke. The federal and some state and local governments have restricted cigarette smoking in public places. At the same time, the federal government encourages tobacco use through price supports for tobacco and a lack of antismoking education.

An opportunity has gone wanting. Organized environmentalists wield much political power. However, they have seldom used this power effectively against tobacco pollution. Tobacco companies are outspending environmentalists through intense print advertising and political lobbying. All Americans suffer from this disparity.

Urban dwellers, in particular, ail from environmentalism's narrow vision. Urban poverty breeds crime, disease, drug addiction, inadequate housing, and a reduced tax base—factors which burden an already deteriorating urban environment.

Many people flee to the suburbs. Urban housing then is left to those unable or unwilling to maintain it. Sometimes properties are abandoned because owners cannot collect enough rent to pay taxes and meet building codes.[8]

These effects of urban poverty hurt even the more fortunate. The costs of fire and police protection increase for everyone. Public sanitation problems escalate. Medical costs force many people to maintain inadequate health care or none at all. Yet, these factors comprise as intimate a part of the urban environment as air does. Some of these problems touch us all.

Sometimes neighborhood clean-up projects brighten the landscape and reduce pest habitat. But slums generally remain slums. "Downtown" still suffers from too much trash and congestion. These aspects of human environment have attracted too little attention from the organized environmental movement. Nuclear power plants and rain forests build more excitement.

The popularity of the environmental movement has lulled people into thinking all the correct issues are being addressed by dedicated professionals. They think the situation is well in hand, and solutions await just around the corner. They have been fooled.

Environmentalists are like the rest of us. They have their pet pollutants, their favorite issues, their ecological idols. We cannot afford to trust others to solve all the environmental problems that plague us. Our conscience needs to move us to improve our world and to think before we act.

We especially should not blindly follow a harbinger of doom. Recycling, for instance, has been pitched as the answer to a solid waste nightmare. But are we really awash in recyclable garbage as we have been told? Studies done by the Garbage Project suggest the magnitude of the solid waste problem has been exaggerated, as have been the benefits of recycling. Not that recycling is useless. We should encourage people to recycle. But we need to encourage effective recycling, not just returning throwaways for the sake of recycling.

William L. Rathje, head of the Garbage Project, used newspapers to illustrate one potential problem with recycling. Very few old newspapers, it seems, are recycled into new newspapers. But recycled newspapers make good wallboard, insulation, cereal boxes, and paperboard for the insides of automobiles. However, the market for these uses is near saturation. State mandated recycling may even worsen this problem. Curbside recycling in New Jersey was blamed for driving down the price of old newsprint from as much as $40 per ton to *minus* $25 per ton. Now, as Mr. Rathje put it, ". . . you have to pay to have it taken away."[9]

The best indicator of effective recycling is still the price paid at recycling centers. The Garbage Project has shown that as the price rises for recyclables, we can expect to find less of these items in local refuse.

All this means is that environmental issues, such as recycling, are subject to economic principles, such as supply and demand. Instead of hating this relationship, we need to use it to encourage demand for products made from recycled goods. But we also need to understand recycling is no cure-all for solid waste problems.

Extremes in environmentalism can cause more harm than simply flooding a saturated recycling market. Extremes can take us where we do not want to go. Environmentalism has taken an ugly turn in recent years. Some people apparently believe violence provides the only remedy to problems that have not been resolved by calmer methods.

A rational environmental conscience not only recognizes a genuine problem, it also recognizes terrorism will never solve the problem. Indeed, militancy has hurt the environmental movement from which it grew.

A part of this problem in the United States may stem from the federal government itself. It owns one-third of America.[10] Public ownership invites conflict over public use.

One hiker in a western national park is afraid he will see a grizzly bear. The hiker just behind him on the trail is afraid he will not. Now there is the quandary. Is a national park for bears or for people? Both hikers probably would give conflicting answers. Each could compromise and find a way to provide a pleasant experience for both. But practical compromise is often ignored. In its place, some people preach sabotage as the solution to environmental problems, making "wise use" of a resource the use these people alone deem appropriate. Radicalism, in their minds, protects the environment.

But radicalism in the environmental arena generates the same distrust it generates in the religious and political arenas. Such radicalism is counterproductive. In most cases, it is also just plain wrong. Illegal or unethical behavior will never remedy environmental problems any more than this behavior will draw people into church. It cannot. It lacks the right answers.

Environmentalism's fervor also has prompted action before the results of that action were fully considered. "Wilderness areas," for instance, have been designated as such in an effort to preserve them in their most pristine state. Human intrusion is expected to be kept to a minimum through a lack of roads, toilet facilities, and other improvements. But officials have learned the wilderness designation acts as a beacon drawing scores more people than the natural situation can accommodate. Pristine loses its meaning when trampled by thousands of human feet.

This is not to suggest we eradicate wilderness areas. But we need to have reasonable expectations for any environmental decision we make. We will find no panacea on earth.

Is environmentalism, then, a curse? Should we cast it aside? Of course not. We need to realize, however, our fervor must be properly directed to solve environmental problems. We need to work to improve the entire web of human environment. Thus, we need to consider matters not always considered by many environmentalists. These matters form much of our day to day environment and intertwine with the more "natural" elements we hear so much about. Humanity is not an intruder. We comprise an intimate part of the earth's environment.

Recognizing our niche within the broad ecosystem we call Earth frees us to become true environmentalists. Otherwise, we distinguish too much between the "natural" environment and that environment built by human hands. We fail to see the results of our creativity and productivity are natural to humankind. A house certainly differs from a bird's nest, but a house is just as natural for a human as a nest is for a bird.

Herein lies the problem with pursuing environmentalism as an end in itself. Humans are considered pollutants. This mindset hinders an effective conscience toward genuine pollution problems.

A conscience relative to environmental pollution may be too restrictive, too permissive, or somewhere in between. An overly strict conscience does not admit we must eliminate modern civilization if we are to eliminate pollution. On the other hand, an overly permissive conscience is easily duped into believing pollution does not exist or is only a minimal problem. Our perspective depends largely on how we define pollution. An ecologist may consider anything that upsets an ecological balance as a pollutant. Consequently, an activity of great human benefit may pollute simply because it alters an ecosystem. The impact of the activity on the resource weighs more heavily than the human benefit of the activity. Thus, ecologists are sometimes thought to be unrealistic. And in truth, many ecologists barely recognize human benefit compared to resource preservation.

An engineer, on the other hand, may be very realistic. To an engineer, pollution may occur only when the imbalance caused by an activity generates an immediate negative impact to people. Engineers carefully plan and design specifically to create direct measurable benefits to people. From this point of view, good design solves pollution problems. However, this viewpoint assumes we know all the important effects of pollution. We do not.

Somewhere between these extremes lies the understanding pollution will occur because much of the earth is crowded with people. People consume resources and produce waste. Resources must come from the environment, and waste must somehow be returned to the environment. Must. Human activity always has and always will alter the ecology of this planet. The pollution question rests on the manner by which this alteration comes and the degree to which an ecological system is affected.

We often blame overpopulation for pollution problems. We suppose we possess the wisdom to determine the carrying capacity of our world—just how many of us are enough. But even more dangerous, we overlook the genuine cause of most environmental troubles: overconsumption.

Few would deny as population increases, resource use also increases. But population size is only part of the picture. In developed countries, in particular, population growth has generated only about one-tenth of the increase in the use of natural resources.[11] The rest of us just keep using more.

We are faced, then, with choices. Consumption can continue to spiral until we run out of some resources, or we can alter our consumption habits. Certainly, none of us want to spoil our home forever. Nor do we want to go without the things we need for our health, safety, and enjoyment of life. But we must make choices. Our consumption habits can change, and we can better manage the resources we are blessed with.

In the past, the costs and benefits of our activities were measured almost solely by money. A dip in the stock market panicked speculators and alarmed the rest of us even if we did not know what a "corn future" was.

We are now learning money worship has put economics and the environment on a collision course. This conflict has no winners. We cannot continue to use resources at greatly increasing rates and not expect to feel the pinch. Many natural resources are not renewable. Even renewable resources must now be renewed from a reduced environment. As a result, conservation attitudes that have often been laughed at are cropping up in more and more people.

They are asking such questions as: Is development the best land-use alternative? Is utilization the best strategy toward a particular resource? Is an increasing gross national product always necessary to our health, safety, and well-being?

Many environmentalists have not admitted the need for hard choices. For these folks, the ecosystem almost always dominates economic needs and sometimes even health and safety needs. Development of any kind may be considered an enemy of the environment.

Sometimes this is the case, hut not always. Humans form part of a naturally occurring ecosystem, and the end result of our activity forms a portion of our environment. The tough part comes in recognizing the point where altering the environment damages it. Not all change is bad. That which is bad needs to be discouraged. That which is not needs to be recognized.

Environmentalism has succeeded as a watchdog over the

"natural" portion of our world. We should be glad for that. But environmentalism has failed to unify all the facets of our environment. This imbalance prevents environmentalism from solving the greatest environmental problem we face: greed. Each of us must solve our own part of that problem. Then, through laws, education, and peer pressure, we all can encourage one another to address the hard choices we now face because of past mistakes.

Our future depends on it.

NOTES TO CHAPTER FOUR

1. Merril Eisenbud, *Environment, Technology, and Health—Human Ecology in Historical Perspective* (New York: New York University Press, 1978), p. 3.
2. Ibid., p. 70.
3. Ibid., p. 61.
4. Ibid., p. 63.
5. Ibid., p. 65.
6. Julie Sullivan, ed., *The American Environment* (New York: H.W. Wilson Co., 1984), p. 34.
7. Eisenbud, *Environment, Technology, and Health*, p. 361.
8. Ibid., p. 362.
9. William L. Rathje, "Rubbish," *The Atlantic* 264, No. 6 (Dec. 1989): 106.
10. Sullivan, *The American Environment*, p. 84.
11. P. Aarne Vesilind and J. Jeffrey Peirce, *Environmental Pollution and Control*, 2nd ed., (Ann Arbor, Michigan: Ann Arbor Science Publishers, 1983), p. 4.

CHAPTER FIVE

✦

ENVIRONMENTAL MORALITY

Are scientists the only ones who have the answers to the problems we face?

Many people assume science holds some mysterious key to our problems, forgetting many of these problems arose from scientific study and development. The controlled use of radioactivity in medical treatment has saved many lives, for instance. The equally controlled use of radioactivity in warfare has cost many lives. Scientists developed both uses.

Perhaps people choose to worship science because they prefer that to accepting moral absolutes. Scientists discover new worlds. Science opens known worlds to new dimensions and proves many old assumptions false. Science is amoral. It allows us to adjust our ethical standards to achieve our priorities without burdening our conscience. This present generation is paying the price of our forbears' abdication of environmental morality. And future generations will pay for our shortsightedness.

Have we forgotten our actions define our character? Naturally, our environment influences how we do things. But our reactions to environmental needs (or any need) issue from an ethical standard, a morality, we have incorporated into our character though the years. As a result, when we perceive that we no longer fit into our environment, our character persuades us either to adapt to the situation or to adapt the situation to us.

Each persuasion has its own supporters. Staunch environmentalists insist the natural situation be favored at all costs. Adamant developers insist nature be altered to fit our immediate wishes. The conflict between the way things are and the way we think they should be can provoke crises.

The fact is, human life-styles prevent us from living in absolute harmony with nature. Crops must be cultivated for food. Trees must be cut for wood and paper. Water must be collected for drinking and irrigation. And land must be used for both homesites and waste disposal. Must. But even though a relationship of peace and tranquility with nature eludes us, we can enjoy a balanced relationship with our environment. Our moral perspective will determine our success.

Humans are the most intelligent and creative creatures that ever lived on earth. Yet, we seem to be the least compatible with the environment. Some people blame this paradox on religion, some on society, some on politics, and some on science and technology. Which, or none, is to blame?

Some modern thinkers believe Western man has caused the greatest environmental damage on earth. Since Judeo-Christian traditions have strongly influenced Western civilization, these thinkers blame Western religions, especially Judaism and Christianity, as the root of environmental problems. They seem to think the divine command in Genesis to subdue nature and fill the earth licenses all manner of environmental abuse. Western man, they imply, considers nature an enemy.[1]

But do religious dogmas mold human morality, or does our morality produce many of our religious tenets? Too often we confuse biblical Christianity with religion. We think the Bible prescribes all the do's and don't's we have formed into religious traditions. This thinking is exactly backward. The Bible fairly shouts religion burdens people with a load no one can bear.

Biblical Christianity, on the other hand, offers each of us a personal relationship with God unfettered by religious dogma. Through this relationship, we can look at our world and its people through eyes tempered with love, humility, and respect. Religion can suppress these characteristics.

But despite the contradictions and inconsistencies of human religions, religion itself never spawned our incompatibility with our environment. Something much deeper had to do that.

Many of those who reject the religion theory of environmental woes blame modern society. Capitalism, especially, is damned for its promotion of individual wealth and self-interest. Indeed, we see much greed and selfishness in capitalist societies. However, these societies also generate the greatest philanthropic gestures we see.

Many who dislike capitalism promote socialism as a superior society. These folks ignore history. Socialist systems, especially Marxism, fail miserably to protect the environment as well as to feed, clothe, and house citizens. Some totalitarian governments have even learned to use pollution as a military weapon. The *Journal of Environmental Health* reported about one such warfare tactic.

During the war in the Persian Gulf early in 1991, Iraqi troops dumped enough crude oil into the Gulf to make the Exxon Valdez spill in Alaska seem minuscule by comparison. This intentional spill threatened several rare or endangered species of plants and animals as well as mangrove stands, sea grass beds, and coral reefs. If tar balls were formed, they would have sunk to the bottom and killed crabs, oysters, and other bottom-dwellers.[2]

Even potable water supplies were threatened. Fresh water in the region is provided through desalination (800,000 tons/day in Jubail, Saudi Arabia). The raw water is pumped from the Persian Gulf. Though Iraq may have viewed the oil release as economic retaliation against the United States, this spill will affect the people of the Persian Gulf region for many years. And cleanup has been estimated to cost $5 billion.[3]

Some say Iraq learned this tactic from the Soviets. A *Los Angeles Times* report referred to Soviet military texts that teach the use of "ecological weapons," such as flooding and pollution, to disrupt navigation and compromise the usefulness of water resources.[4] Totalitarianism does not encourage environmental concern.

Today it seems only primitive societies live in balance with their environment. But even primitive societies can greatly alter their environment, and not always for their good. Yet, we need not blame society for environmental troubles. Societies grew out of the needs and wishes of people and reflect the character of the people making up those societies.

Since societies are political in nature, we might conclude politics prevents us from living compatibly with our environment as well as with one another. A great array of political systems populate the globe, and their conflicting ideologies prevent agreement on any issue. Environmental issues have taken on a strongly political aura so that environmental managers must consider the political ramifications of almost every decision they make. Consequently, we often blame world politics, especially free world politics, for the mess we have made on our planet.

We sometimes forget people produce political parties and regimes. Like society in general, political systems grow out of the needs and wishes of people. We can disavow support for a particular system. We can, and should, deplore the corruption we see in virtually every government in the world. But if we are to recognize the cause of environmental problems, we must understand our politics, like our societies, reflect our character. Politics do not pollute; people do.

Our problem is that we do not want to account for our behavior. We think the other guy deserves the blame. Thus, politics and politicians provide convenient scapegoats for any environmental crisis we identify.

But politics rises from civilization. Even primitive societies formed political structures, and some of these societies exploited their environment. Did tribal politics lead the Maori to exterminate the moa, a large flightless bird, in New Zealand?

We might better blame hunger and lack of foresight. No mammals (except a bat) inhabited New Zealand back then, and the moa was the Maori's only source of meat.[5] Apparently, the Maori did not think they could exhaust their food supply.

We are left, perhaps, to impugn guilt for environmental ills to science and technology. Science gave us the nuclear bomb, for instance, and technology can deliver the bomb to any spot in the world. In addition, modern polluting industries grew out of the great technological strides of the Industrial Revolution.

However, at least two difficulties exist in blaming science and technology for environmental problems. Primitive societies, like advanced societies, often degrade their world. And modern science and technology benefit the whole world.

Knowledge is neither good nor evil. We display our goals and motives, good or bad, in our application of knowledge. By examining ourselves——why and how we do things—we can identify the true culprit in our environmental crises.

Why, then, do we spoil the only planet capable of sustaining us? For the same reason we destroy one another. In James' letter to the Church, he asked, "What is causing the quarrels and fights among you?" (James 4:1a, LB).

His answer vividly describes the logical and expected outcome of human greed. "Isn't it because there is a whole army of evil desires within you? You want what you don't have, so you kill to get it. You long for what others have, and can't afford it, so you start a fight to take it away from them. And yet the reason you don't have what you want is that you don't ask God for it. And even when you do ask you don't get it because your whole aim is wrong—you want only what will give *you* pleasure" (James 4:1b-3, LB).

We humans display an excessive desire to possess wealth far beyond what we need or deserve. We all understand how greed drives industry executives and professional athletes to demand salaries unimaginable to most of us. But greed also leads developers to fill wetlands for more condominiums when nearby condos are failing for lack of sales. Greed compels one body of consumers to demand water at the expense of other consumers. Many hunters and fishermen selfishly ignore game and fish laws enacted to perpetuate some of our natural wealth. Industry pollutes as much as the law allows because reducing polluting emissions reduces profits. And governments sometimes circumvent environmental laws because officials stand to gain by it.

But greed flaunts a vicious cycle. Greed reproduces. It grows, even feeding on itself, to twist our conscience into a self-centered moral ethic. Greed justifies all manner of abusive practices while blinding us to our own selfish motives. It is no wonder the Bible condemns our love of money as the root of all manner of evil. Wealth never satisfies. We always want more.

Jesus taught a parable about a man who bears a striking resemblance to many people today (ref. Luke 12:16-21). This man possessed much more wealth than he needed for his and his family's health and well-being. He managed his affairs well, for his land remained very productive year after year. Eventually, he realized he had a problem. He had accumulated more assets than he could use and even more than he could store. His solution? Build bigger buildings to accommodate more. This man thought hoarding would secure his future. Soon, though, he learned he could not take it with him.

Jesus did not condemn wealth with this parable. He pointed out the foolishness of a life-style motivated by greed. Herein lie the environmental problems we face. Striving for material wealth, we impoverish our planet. We maximize profits at the expense of our neighbors and of future generations. We industrialize more and more and intensify sordidness and squalor. We trade life for riches.

Each of us loses in this struggle. Great wealth insulates those who possess it from many of the economic and environmental factors that plague the rest of us. But polluted air chokes us all.

Why, then, do we all not jump on the environmental bandwagon? Why do industries not compete in the arena of environmental protection as much as in the production and marketing arenas? Why do many hunters and fishermen not admit bag and size limits exist for good reason? Why?

The answer breaks down into at least two components. The first: egoism bases our behavior on self-interest. The second: we do not think nature has value apart from its direct utility to humans.

Looking at our environment through the smudged lenses of self-interest blocks us from forming sound environmental ethics. And ethics, without an absolute standard, bend with the everchanging wind of public opinion. Thus, we find ourselves valuing life itself according to the escalating standard of materialism. How much is enough?

Ethical theories were developed primarily to resolve conflicts among people. Even environmental ethics have most often based values on human utility. Environmental problems arise among people because what one person does affects others. But environmental conflicts among people began as conflicts between people and nature.

Sometimes our goals bump up against environmental systems already in place. Then one or the other has to yield. Manipulating nature to achieve our goals is not necessarily wrong. Neither is it necessarily right. Thus, we may have a moral conflict and need to make a choice. Complicating this choice are two inequalities we insert into the equation, both arbitrary.

We insert a human inequality because we consider some people to be worth more than others. Likewise, we judge some elements of our environment to be worth more than other elements. We rank human worth primarily by financial success and environmental worth primarily by market value. From such thinking, we have devised the hellish concept of situation ethics.

Many boaters have decided their freedom to drive fast is worth more than the few manatees remaining in Florida. Numerous studies have shown injury from power boats threatens the manatees' survival more than any other human factor. We admit manatees have value, but only as long as they do not interfere with our boating.

A situational appraisal of the environment causes environmental crises in the same way a situational appraisal of people causes human crises. We consider only part of the picture, usually money. But appraising nature correctly requires an understanding of how a natural system is valuable. Two viewpoints commonly crop up.

The instrumental view assigns value based on human utility. In this view, nature possesses no real value of itself but is valuable to the extent it can be used by people. A strictly instrumental view prompts people to exploit their world.

In many cases, this viewpoint led people to eradicate natural systems because they hindered the desired use of a piece of land, such as for a shopping mall. In other situations, this view led to the development of conservation practices that prolong the value of a natural system, such as a forest. An instrumental view of our environment considers the relative values of different resource uses and normally selects the most profitable value.

An intrinsic view recognizes value in nature apart from any monetary return. The natural element possesses a value simply because it exists. We cannot consume abstract values, and we find them difficult to tax. We must assume intrinsic value somewhat by faith. We believe in it.

Adhering to either viewpoint alone prevents us from resolving environmental conflicts. On the one hand, we may eliminate a resource's existence. On the other hand, we may prevent a resource's use. Neither extreme produces effective environmental policy.

We need to marry the intrinsic and instrumental views of nature to resolve most environmental conflicts. Admittedly, this task requires faith. We need to believe nature possesses value apart from monetary return.

We also need to identify nature's value to humanity. This point of view can encompass financial value, as with a housing development, but it also can include less tangible benefits. The value of clean air, for instance, cannot always be measured by money. And some of us just feel better knowing bald eagles still soar over North America.

If we divorce intrinsic and instrumental values, we risk setting our environmental morality too low. We could even revert to the attitude that led some Old World Europeans to think New World Indians were subhuman. We cannot view our environment with an enlightened conscience if we think part of that environment is worthless. Nor can we understand our own ecological value unless we recognize our interdependence with our environment. Everything has its place.

Still, the intrinsic value of nature is hard to assess. We might find it easier if we quit trying to assign dollar values to intangible worth, such as aesthetics. This habit will be hard to break. But if we do not at least moderate it, we will leave ourselves with two options. We can try to withdraw from any environmental impacts, which is impossible. We would have to remove ourselves from the environment. Or we can try to eliminate that part of the environment not readily used by people.

We have employed this second option during development "boom" times and now face the consequences of ignoring the natural value inherent in the environment. In some cases, we have seared our conscience to any worth other than financial.

However, we can train our conscience to recognize the environment's intrinsic value and at the same time appreciate the human benefits of a healthy environment. But we will fail as long as we remain environmentally ignorant. Environmental crises surmount our emotions, and effective action extends far beyond saving whales and cleaning industrial emissions.

Each person forms an integral part of the environment. Thus, each of us needs to live an educated and sensitive environmental ethic. To accomplish this task requires a sense of morality since we have responsibilities toward one another and toward our world. Until we accept that fact, we will never truly protect ourselves and our habitation.

NOTES TO CHAPTER FIVE

1. P. Aarne Vesilind and J. Jeffrey Peirce, *Environmental Pollution and Control*, 2nd ed. (Ann Arbor, Michigan: Ann Arbor Science Publishers, 1983), p. 361-362.
2. Ginger L. Gist, "The New 'Dead' Sea," *Journal of Environmental Health* (Spring 1991): 21-22.
3. Ibid.
4. *Tampa* (Fla.) *Tribune*, 29 January, 1991.
5. Vesilind and Peirce, *Environmental Pollution*, p. 365.

Chapter Six

A Balanced Conscience

Should we log forests or save woodpeckers?

If you answered too quickly either way, you may have signaled a lack of one of the most essential ingredients for a healthy environmental conscience: balance. Wildlife warrants protection. People, however, have needs just as critical as any animal's.

In 1990, the United States Forest Service began a population survey of the red-cockaded woodpecker. These woodpeckers inhabit old-growth pine forests in the Deep South. Not just any pine will do for a nest. These birds depend on living pines seventy years old or older that are infected with the fungus that causes red heart disease. This disease rots the interior of very old pines, enabling these woodpeckers to chisel out a nest. Younger, healthy pines are rarely used. Without human help, this woodpecker's strict habitat requirements may spell its doom.

Early in 1991, only about 4,000 colonies of these birds were thought to exist throughout the Southeast, mostly in national forests. But thinking of normal colonies misleads us. A red-cockaded woodpecker colony contains no more than one breeding pair of birds, though several younger birds may be included. In the Apalachicola National Forest in northern Florida, 684 colonies were reported, but nearly a quarter contained only single males.[1]

In 1991, the Forest Service temporarily banned clear-cutting of pines within a three-quarter mile radius of each colony of red-cockaded woodpeckers in the Apalachicola National Forest.[2] With this ban they hoped to give the woodpeckers some breathing room. The birds need the old trees to build nests. They also need time to mature. Equally important, the Forest Service needed time for a long-term recovery program for these woodpeckers to clear the maze of bureaucratic approval. Continued clear-cutting of old-growth pines would not have allowed the woodpeckers or the Forest Service the time each needed.

Conservationists praised the partial ban on clear-cutting while timber interests denounced it. This conflict illustrates a common problem with environmental issues. We are forced to take sides.

Competition is fine in sports, but we are not playing a game. We all depend on a healthy environment for our very lives, and conserving that environment is a deadly serious business. We all could lose in such a conflict.

Cooperation will end the battle over the environment. We will not cooperate, however, until we balance our environmental conscience. To achieve this balance, we must consider the inherent value of each component of the environment, and at the same time consider the public and private interests in those components.

It will do us no good to chastise all development. We all like to drink clean water and to flush away our wastes. We like to buy manufactured goods and to sleep under a roof. To benefit from our environment, we sometimes must alter it. To benefit the *most* from our environment, we must not destroy it. Sometimes we need to adjust to things as they are.

We need wisdom. For some reason, we think wisdom is a

mysterious faculty only some people possess. And, of course, the wisest people agree with us. Wisdom, however, is born of careful thought and experience. When we hastily choose sides, without careful consideration of alternative views, we just might make the right choice. Too often, though, our haste displays a characteristic not so wise.

The Bible puts it this way: "Professing themselves to be wise, they became fools" (Rom. 1:22). In this passage, the apostle Paul was describing the outcome of the Romans' idolatry. Having rejected their Creator even though he was displayed clearly in the Romans' environment, they chose to worship inanimate replicas of birds and beasts. Paul noted they had exchanged God's glory for the "wisdom" of worshipping man-made idols. "How foolish," he declared.

Paul could just as easily have noted our foolishness of rejecting the inherent worth and long-term benefits of a natural resource in favor of an immediate profit. He also could have warned us of the foolishness of opposing all development just because it alters the environment.

But here the similarity ends. There is no middle ground in our relationship with God. We accept him or reject him. Environmental issues, however, often possess a middle ground begging for wisdom to see ways to improve our lives while protecting the integrity of our world. In this element of environmental issues, a balanced conscience will shine.

Wisdom rarely, if ever, is displayed in arrogance. Arrogance prevents us from seeing an alternative, from recognizing a mistake. There simply is not time. Arrogance drives us to push for our own interest and to dominate. For this reason, an arrogant attitude will never develop into a balanced conscience. Balance demands humility.

Secular society rarely teaches humility. Most often it preaches determination and drive. Both qualities help us achieve our goals. But identifying goals worthy of our energy and resources requires humility.

Humility also helps us avoid some mistakes by revealing different sides of the same issue. We will never avoid mistakes altogether. But admitting we do not know all the answers and that maybe our "opponent" knows something reduces our margin of error.

Many people confuse humility with self-depreciation. In order to be humble, they think they must debase themselves. They could hardly be more wrong.

Modesty recognizes we are intelligent beings, but at the same time do not know all there is to know about every issue we face. Each of us knows part of the truth. By pulling together rather than pushing apart, we can see more clearly the complex web of life of which we are only a part.

People who disagree with us may still have a valid point. Recognizing that point requires wisdom and humility. Acting on that point requires deference. If we never find strength to yield to another's view or need, we will never see our environment in a whole sense. The world, our nation, and our state will exist primarily for our own benefit. We will view others as lesser team mates if they agree with us and as usurpers if they do not. More than likely, we will forfeit our effectiveness in the process.

The key is to know when to defer. Sometimes we wrap ourselves so tightly in our own perspective we fail to see any validity in another point of view. But if we look at the real issue, and if we recognize we may learn from someone else, then we usually will know when to bow and when to stand.

Wisdom. Humility. Deference. These three keys to a balanced conscience must be practiced to be effective. We need to "do," but we need to know what to do and when. Learning this takes time.

Time, however, can overwhelm us. We need to overcome the habit of waiting until a crisis occurs to address an issue. Had we chosen to protect red-cockaded woodpecker colonies a few years earlier, these birds might not inhabit such a precarious niche today. We may even now have waited too long to turn their situation around.

Reversing our habit of putting off environmental decisions until the last minute requires a radical change in our conscience.

Our traditional view of the world has reflected various cultural, religious, and personal outlooks. Some societies considered natural resources to be the property of a select few, such as the feudal lords of medieval Europe and Japan. These people practiced rudimentary resource management, to be sure. However, the general populace gained little from these practices. Early American Indians claimed possession, though perhaps not ownership, of vaguely defined territories. Competing bands fought to maintain control of their traditional territories and to gain new grounds. Minor resource management may have been practiced inadvertently when prairies were burned over to drive game. But the goal was strictly utilitarian.

Following independence from Britain, Americans developed the awesome concept of Manifest Destiny. God, they believed, ordained Americans to control the vast domain from the Atlantic Ocean to the Pacific. Indians and Mexicans were evicted from lands they had occupied for generations, and resources were claimed by individuals of all ranks. Capitalism recognized the value of a resource primarily in its utility.

More recently, Communism reduced the worth of individuals to such low measure that production was made the only goal of society. The environment, and even human life, ranked way down the list. It should not surprise anyone that environmental problems in the former Soviet countries have been more severe than in the West.

But society and politics do not corner the market on environmental ignorance. Hindus in India worship their animals and starve themselves. Christians in the Free World all too often plunder their environment while trying to "subdue" it. Religion, it seems, treats nature either as a god or as an enemy.

Communism or capitalism. Democracy or dictatorship. Religion or politics. The notions of men have failed to give us the impetus both to use and to protect the environment. Our failure falls squarely at the feet of our greed. We jealously hoard what we have. We crave what we do not have and try to take it by force. We end up fighting ourselves and each other.

Only one concept combines wisdom, humility, and deference into a lifestyle that protects the integrity of the environment for our present benefit and for the benefit of future generations. This concept, stewardship, is dynamic enough to enable us truly to conserve the environment we live in.

Stewardship precludes selfishness. At the same time, it considers the needs unique both to private ownership and to public good. However, modern people have generally maligned stewardship as an outmoded notion of religious faith. They apparently believe stewardship confines us to a "garden mentality" that disallows the application of science and the recognition of individual ownership. To them, stewardship so intertwines with Judeo-Christian tradition, those who reject that tradition also reject the concept of stewardship.

It is no wonder. Too often we church folk relate stewardship only to money. This emphasis waters down the much broader implications of the word.

Stewardship means delegated authority over another's possessions and can apply equally to people or to property. The biblical thought is one of directed privilege. We have been given the privilege of overseeing God's Creation.

Many people vehemently oppose any attitude or plan that hints of a religious perspective. In the biblical concept, stewardship implies administration under divine direction. But stewards need not recognize a transcendent God or believe in future rewards in another life. It would be nice if they did. However, stewards need only recognize the world exists for no one person and for no one generation. Other people need what we possess, and this need will pass to future generations, too.

Stewardship compels us to recognize the inherent worth of a resource apart from its manipulation. A steward's conscience frees us to live contentedly within our ecological niche while benefiting from our rather exalted position on earth. Thus, we can find a balance in our relationship with our environment.

The problem is that those who do not acknowledge God find it almost impossible to maintain a steward's conscience. It is tough enough for believers. Since stewardship requires a sense of being entrusted with something, keeping this attitude is easier if we have someone to trust. Who will we trust ultimately? Ourselves? Money? Government? Not if we are wise.

Our growing population lives on a finite planet with declining resources. If these resources are to continue to provide for our health, safety, and welfare, we must act as stewards of these resources. To act otherwise will surely bring want beyond anything we now know.

Usually, we either place nature on a pedestal for protection from human use, or we reject any value of nature apart from its utility. On the one hand, we hoard the environment. On the other hand, we plunder it.

We might think hoarding is more desirable than plundering if we do not see hoarding's outcome. Remember the parable mentioned in Chapter Five? A rich man managed his land so well it yielded far more crops than he could use or store. Did he sell the excess? Give it to the needy?

He decided, "This is what I will do: I will tear down my barns and build larger ones, and there I will store all my grain and my goods. And I will say to my soul, 'Soul, you have many goods laid up for many years to come; take your ease, eat, drink, and be merry'" (Luke 12:18-19, NASB).

Sadly for this man, God judged his greed that very night: "... now who will own what you have prepared?" (Luke 12:20b, NASB).

Hoarding a portion of the environment evolves into covetousness that restricts more and more the availability of that portion. We can easily lose sight of our relationship with the environment. The environment at that point is shut away from us rather than shared among us and conserved.

When we hoard a resource, we eradicate its usefulness. When we plunder a resource, we also eradicate its usefulness. Stewards learn to live with their environment. Environmental balance becomes their lifestyle.

We need not insist, however, that preservation has no place in the management of our environment. In some cases, preservation offers the only means to protect certain elements of the environment. The forest resource needed to sustain a red cockaded woodpecker colony must be preserved if we are to help this species survive. Similar situations exist throughout the world. But we will not see preservation as a tool for resource conservation if we reject the inherent value of that resource.

We do not eat woodpeckers. Still, many of us find pleasure in knowing they exist and in seeing them occasionally. And many of us know woodpeckers play an important role in the ecology of this planet which includes humans. Thus, preservation can benefit us all.

Nevertheless, conservation implies use. Management techniques have been developed to increase resource production and the efficiency of resource use. Traditionally, we have geared our environmental efforts toward active use. Bobwhite quail management, for instance, developed in the face of a declining population of a favorite game bird. Habitat manipulation provided the key to the quails' survival and growth.

In developed nations today, greater emphasis is being placed on passive uses of the environment. Non-game wildlife receive attention where once only game species were thought valuable enough to conserve. Recreational use of public lands receives more support today than in past generations. And the quality of a resource, such as water, is regulated as much as its quantity. All this should encourage us. Our conscience is changing.

But we still hold on to misplaced emphases. Development, though we need much of it, still degrades the environment enough to reduce its long-term value. Many people still think their own immediate desires are more important than anyone else's needs. And too many of us still think our environment can forever absorb the abuse we hurl at it without faltering. We ignore the evidence to the contrary.

We are tempted to blame all our environmental problems on overpopulation. We think there are too many of us to share the world's resources. This just is not true. People adapt and innovate. We produce more from less. There are not too many of us to share. There are too many of us unwilling to share.

This is not to say all our resources are infinite. They are not. But nature drives to perpetuate itself. Unless interrupted by some natural cataclysm or by relentless human abuse, our environment will keep producing. We can help it produce more. We can "fill" the earth and still maintain the earth's environmental integrity. Up to this point, though, we have only loaded our planet with people largely motivated by greed and sometimes hate. Our environment has been used as a tool by some to get rich and by others to get even.

The psalmist wondered, "Why do the nations rage and the peoples plot in vain?" (Ps. 2:1, NIV). The prophet Isaiah asked, "Why spend money on what is not bread, and your labor on what does not satisfy?" (Is. 55:2, NIV).

How desperately we need to know what works. Stewardship does not preclude private ownership and free enterprise. Environmentally sound farming practices benefit the farmer and the consumer. Advanced waste water treatment helps preserve the quality of life for everyone. Careful urban planning reduces the human and environmental stresses of congestion and pollution. A conscience sensitive to the needs of our environment and to the needs of its people, a steward's conscience, enables us to be more productive, less destructive, and better satisfied.

For some reason, we think accountability to a transcendent Sovereign precludes sound judgement, scientific inquiry, and practical common sense. Nothing could be farther from the truth. Our need of good judgement, good science, and good sense means we cannot escape accountability; we can only ignore it.

Today we face the economic and environmental consequences of ignoring accountability for many of our decisions over the past century or so. We have chosen to blame overpopulation rather than the greed of that population for our problems. Thus, long-term solutions have eluded us.

The solution to environmental problems, or to any other problem, is not to kill babies or elderly people. The solution is to live responsibly toward ourselves, each other, and our environment. No one person can reshape a national or even regional conscience. But we can shape our own.

We can find alternatives to current resource use. We can alter our life-styles to reduce our use of a resource. We can produce more from what we have. By doing so, we will learn to live with our environment without plundering it. We can develop a balanced environmental conscience both in the world view and in our own little part of the world. Then we will know how much is enough.

NOTES TO CHAPTER SIX

1. *Bradenton* (Fla.) *Herald*, 20 May, 1991.
2. Ibid.

CHAPTER SEVEN

✦

WHAT ABOUT THE THIRD WORLD?

"IN MOST OF THE THIRD WORLD A SLOPPING-AND-
SCAVENGING SYSTEM (OF GARBAGE DISPOSAL)
THAT HECTOR AND AENEAS WOULD RECOGNIZE
REMAINS IN PLACE."[1]

WILLIAM L. RATHJE
UNIVERSITY OF ARIZONA

Major industrial nations today lead in the war against environmental abuse. It is a different matter in the Third World.

All nations share such problems as air and water pollution and natural disasters. But Third World nations face extremes of such plights as deforestation, desertification, animal and plant extinction, marketing endangered wildlife, toxic waste shipping from industrialized nations, and the disposal of garbage and human waste.[2]

These problems tell us a lot. Not only do these people

suffer from the harsh extremes of nature, they also suffer from their own exploitation of resources. And if that is not enough, they suffer environmental burdens dumped on them by the rest of us.

These problems tell us something else, too. Third World environmental problems turn on economics as tightly as do similar problems of industrialized nations. But the Third World experiences a special dimension of economic pressure. Much of the Third World's population fights merely to survive. Families struggling in abject poverty rarely consider the environmental impacts of overgrazing, slash-and-burn farming, and coral reef poisoning. They must eat—today.

Most of us can understand impoverished nations favoring economic development over environmental integrity. Most of us like to eat, and most people dream of improving their situation. There is no harm in that. The problem arises when we forget there is a limit to what we need for our lives and even for our comfort and convenience. Anything more is wasted on us. This fact holds true in Third World nations as well as in the industrialized nations. But lack of adequate food and water supplies in some Third World countries complicates environmental issues by mingling need with greed.

Just as in more affluent nations, environmental damage can counter long-term benefits of economic development in the Third World. It seems newly industrialized nations especially risk this danger. The extensive petrochemical industry in Taiwan, for instance, has boosted the Taiwanese economy. But this industry also has polluted the air and contaminated fishing waters.[3]

The people of Taiwan may see the day when environmental degradation robs them of some of the benefits of an improved economy. They also may watch economic improvement flee Taiwan for countries with a more sound environmental base. Economic development without adequate environmental controls gambles the future for the present. Sooner or later, someone will lose.

Often we think pollution must accompany economic development. It is no wonder. Everywhere we look in the industrial world, we see the two together. But it does not always have to be so. If the two outcomes *had* to occur together, our task would be simple: preserve underdeveloped regions in their underdeveloped state. Natural resources would rarely be squandered, and plants and animals would rarely be endangered. Clean air and water would rarely be polluted.

The fact is, though, some of the world's worst environmental problems occur in underdeveloped regions many miles from population centers. Many threatened and endangered species, for example, struggle to survive in the Third World: northern white rhinos in Zaire; Coquerel's dwarf lemurs in Madagascar; flightless cormorants in the Galapagos Islands; and wooly spider monkeys in Brazil. If we accept Communist China as an underdeveloped nation, we can add to the list everyone's favorite teddy bear, the giant panda. And there are many more.

Many people deplore clear-cutting in forests of the United States, but deforestation is a serious problem in much of the Third World. Thick forests once blanketed Haiti, for instance. Disappearing as early as the late 1800's, forests covered less than five percent of Haiti's land area by 1987.[4]

The forests were cleared primarily to grow food crops. Deforestation, however, accelerated soil erosion and reduced water available for irrigation. Thus, deforestation was counterproductive. Once a productive agricultural region, today Haiti must import much of its food.[5]

The rain forest of the Amazon region in Brazil may be disappearing at a rate of about 13,000 square kilometers (over 5,000 square miles) per year.[6] Slash-and-burn farming exposes the thin topsoil to the pounding rain, which erodes the soil and soon robs the land of its usefulness for agriculture. In a very few years, farmers must clear more forest. The net result: loss of soil, loss of forest, and loss of productivity.

If deforestation in Haiti and Brazil affected only those countries, the rest of the world *might* afford a hands-off stance. But large scale ecological destruction affects us all. Brazil's rain forests influence air quality outside Brazil. People in other countries now must feed the Haitians. We no longer can afford the attitude that insists "what happens over there is their problem." More and more, it becomes our problem, too.

Can we, then, encourage environmental responsibility in countries where so many people barely survive?

No single solution exists to environmental problems in Third World nations. Yet, the same principle that helps resolve such problems in industrialized nations can help in the Third World. We need to balance the economy and the environment.

So far we have found this task easier said than done. We think we can just send money. All we have done, though, is weaken our own economy without helping theirs. Foreign Aid, as we usually practice it, is not the answer.

One reason, perhaps, is because the whole world competes for resources. In much of the Third World, poverty fuels a particularly intense competition between the economy and the environment.

Regarding such conflict, writers in the *Journal of Environmental Health* concluded, "these competing interests must be approached in a balanced manner, so that short term economic benefits do not lead to long-term economic and environmental catastrophe."[7]

Balancing the economy and the environment is possible. The great strides we have made in renewable resource management attest to this fact. And in recent years, emphases for many of us have changed.

The United States, in particular, grew through a period of wanton exploitation of its environment. We thought our natural resources were limitless. In our first century or so, Americans trapped beavers and hunted buffaloes until they almost disappeared. Farmers scalped and eroded the southern Piedmont to grow more cotton, and industries dumped their wastes into northern lakes and rivers. Miners scoured and pitted western hillsides hunting gold.

Through it all, we laughed at critics who thought we might be destroying the very resources we were exploiting. In the process, we built a nation. But in our wake, we left a wounded environment and a broken and diseased native population.

In recent years, we have recognized much of the damage we caused developing the United States. We have called for more rigid control of the activities we formerly embraced in a freewheeling fashion. More importantly, we have studied. We have learned there is more to successful farming than plowing and planting. We now know conserving forests involves more than cutting trees. We are learning virtually everything we do carries with it an environmental price tag.

If we smoke, we pollute the air for someone else. If we build a housing development, we remove that land from resource production. When we crowd together in cities, we create congestion and waste water problems. There is always a trade-off.

This fact applies equally to the Third World. There rain forest is traded for a few short years of crop and cattle production. Clean water is traded for open sewers. The list goes on. But in the Third World, the distinction between the "haves" and "have nots" glares more blatantly. And in many cases, a destructive environmental conscience is stimulated from abroad.

The answer to environmental problems in the Third World is the same as for the industrialized world. We need to balance the economy and the environment. To succeed, individual and corporate consciences must change. Once that change is accomplished, people can address the specific issues that so far have sharply divided the economy and the environment.

Environmental responsibility is possible throughout the world. The quality of our future depends on it.

NOTES TO CHAPTER SEVEN

1. William L. Rathje, "Rubbish," *The Atlantic* 264, No. 6 (Dec. 1989): 100.
2. Lewis A. Mennerick and Mehrangiz Najafizadeh, "Third World Environmental Health," *Journal of Environmental Health* (Spring 1991): 24.

3. Ibid., p. 25.

4. Charles E. Cobb, "Haiti: Against All Odds," *National Geographic* 172, No. 5 (Nov. 1987): 648.

5. Ibid.

6. Priit J. Vesilind, "Brazil: The Promise and Pain," *National Geographic* 171, No. 3 (March 1987): 382.

7. Mennerick and Najafizadeh, "Third World Environmental Health," p. 24.

CHAPTER EIGHT

✦

GOVERNMENT'S ROLE

M ost of us probably would agree "Big Brother" grows bigger by the day.

Government's tentacles reach today into areas formerly off limits to public oversight. We often lament this fact for good reason. Government regulators can display an unbelievable lack of common sense, much less compassion, in their zeal to enforce rules they may not even fully understand. Then intent of the law bows to legalism.

In 1989 in southern Florida, creative "farmers" dug up part of a forested wetland and piled the sediments to form exposed mounds. Then they planted marijuana on the mounds. When the sheriff found the site, he destroyed the marijuana and the mounds, restoring the integrity of the wetland as well as doing his job.

But someone complained. Florida's wetlands rules required state permits for dredging or filling in certain wetlands, including this one. State environmental regulators penalized the Sheriff's Office. It did not matter that a crime was stopped and a wetland was restored. There was no permit.

Incidents like this one inflame a public already suspicious of government interference. Yet, we need government regulation. Many people show little moral concern for the effects their actions have on others. Where morality falters, law needs to protect.

We generally recognize today that intoxicated drivers hurt us all. Innocent people die. Insurance rates soar even for non-drinkers. Productivity declines when workers cannot work due to injuries received in an accident. The public has learned the need for strict laws governing drinking and driving. These laws benefit us all.

The same principle holds true with the environment. Our lifestyles largely determine soil stability, air and water quality, fish and wildlife production, and on and on. However, the effects of our lifestyles may appear elsewhere and to other people.

We humans are prideful and self-focused. Our nature prompts us to look out for ourselves even at the expense of others. Thus, government needs to guard the interests of those we do not consider. And each of us is not considered by someone.

Most of us accept, though sometimes grudgingly, that we need a degree of oversight. We have developed laws reflecting the changing needs of an increasingly industrialized and rapidly growing society. But law has not always kept pace with society.

Common law developed as judges decided individual cases. People damaged by pollution could take the alleged polluters to court and seek relief through the cessation of polluting activity and/or the payment of damages. Earlier court decisions became precedents. Similar cases, in theory, were decided similarly.[1]

Common law forms the basis of American legal tradition and has served us well for generations. But common law contains a flaw relative to environmental issues. Environmental common law has often turned upon the doctrines of ownership and prior appropriations.

In the first, property ownership decides the issue. A property owner may be allowed to create an open dump because he owns the property.

In the second, reasonable use of a resource may be determined on a first-come first-served basis. A factory may be allowed to dump its waste water into a river because the factory was there before anyone else.

In such situations, common law alone seldom addresses the effect of pollution on the quality of a resource, except as it regards a nuisance. But a nuisance is often the least important aspect of environmental pollution.

We should not abandon common law. But due to the difficulty of addressing pollution through common law, Congress and state governments passed laws specifically related to environmental matters. These statutory laws led to a host of standards by which to judge environmental quality. These laws also provided means to regulate activities affecting environmental quality.

Some activities, such as waste water discharges and industrial emissions, are restricted through effluent and emission standards. Other activities, such as dredging and filling in wetlands, are regulated through permits. Some of these and other activities are governed by both permits and standards.

This system does not eliminate environmental degradation but restricts practices that tend to degrade the environment. A factory may meet effluent standards and permit conditions and still degrade water quality in a small stream. We can improve the situation by setting minimum quality standards for receiving waters, but laws alone do not stop pollution.

Nevertheless, statutory regulation provides two strong forces for conservation. This regulation replaces one-at-a-time judicial proceedings with prescribed standards. It also reduces the inconsistent nature of environmental enforcement. Statutes do not eliminate the tough choices we all need to make in supporting reasonable standards of ambient air and water quality. But statutes reflect government leadership.

History shows that governments must display disciplined leadership if those governed are to practice similar discipline. However, governments have sometimes led the rest of us in degrading the environment. Navies dump their garbage in the ocean. Local governments complain about, and sometimes circumvent, state permitting requirements. Congress and the President confuse us by proclaiming environmental crises on the one hand and encouraging oil exploration in fragile coastal ecosystems on the other. When our government gets away with something, we want to do it, too.

Governments play an indispensable role in environmental conservation. Governments see the "big picture." Individuals and even organizations tend to look at an issue with blinders. They see only the component of the issue that affects them.

Developers, tax assessors, and property owners may look at only the economic aspect of a project. Environmentalists may see anything less than strict preservation as a sell-out. But governments can consider the economic, public health, and environmental impacts from human activities. Most of us have proved poor watchdogs over the environment entrusted to us. Government can fill the gap.

Free World governments, in particular, need to lead the world in the wise management of the environment. We need not expect totalitarian governments to place a high value on the environment. These governments do not even place a high value on life. Repressive governments and shaky economies do not lend themselves to sound resource management and effective pollution control. Therefore, the rest of us bear a greater responsibility, like it or not.

At least in theory, governments represent the interests of the people residing in the various countries, states, and cities. Totalitarian governments, of course, deviate from this theory. Even so, all governments have a responsibility to help conserve the environment for the good of their citizens.

Exhaustive research has taught us how to manage forests and range lands for increased resource production and multiple uses. Research has yielded management techniques that have improved some fish and wildlife populations in the face of increased development. But in spite of successes in managing much of our natural environment, we struggle with our failure to manage the environment created by humans.

In some cases, we are ignorant. It seems we do not yet

know how to solve the crime problem exploding around the world. We need to study. Many times, though, we know what to do but lack the will to do it.

Self-interest encourages us to degrade and devalue our world because so many of us do not recognize values other than money. This shortcoming magnifies the role government needs to play in conserving our environment. Individually, we do not do the job.

Governments need to enact appropriate environmental laws. We have experienced the weakness of using common law alone to handle pollution and land-use problems. We need legal standards by which to judge environmental quality. Early efforts, such as the Clean Air Act of 1963, did little to clean up America's environment. But since the 1970's, Congress has put some teeth into federal environmental regulation, and states have followed its lead.

In spite of establishing standards of environmental quality, governments sometimes fail in their broader responsibilities. Laws alone do not protect much.

Research determines reasonable standards and improved management strategies. Permitting restricts activities which degrade the environment and forces us to address issues we might rather ignore. Monitoring shows us if we are meeting standards and following permit conditions. Enforcement encourages us to follow the rules and provides for corrective measures when we do not. Education informs us of the best conservation measures and why they are important. And a good example gives us a leader to follow.

RESEARCH

Research is study. We study to learn effective and achievable standards of environmental quality. We also study to learn the best ways to maintain these standards. But research may provide its most valuable service inadvertently. It slows us down.

Many coastal areas of the United States experienced "boom times" of development when estuaries and wetlands were destroyed to build subdivisions and shopping centers. People paid little heed to the habitat being lost and to the water quality being degraded. We were making lots of money but did not realize at what cost. Had we studied our environment more and heeded our findings, we almost certainly would enjoy now a cleaner, more healthful environment. Today, even with increasing development pressure, we are slowing down a little to learn about our impacts on the environment and ways to minimize those impacts. Research is the key.

PERMITTING

Permits also slow us down. Applicants may resent this, but permits force us to consider some of the lessons we have learned through research.

We know wetland destruction affects people far-removed from any wetland because of the cumulative effects of regional and national wetland loss. We also have learned alternatives to some of the most destructive dredge and fill practices, such as vertical sea walls. Permitting requirements make us examine those alternatives.

On the negative side, permits do not prevent environmental impacts. And permits increase the cost of development. But permits regulate impacts, and for that reason, we need to support reasonable permitting requirements.

MONITORING

Without effective monitoring, standards and permits mean little. We monitor our environment primarily to provide two measurements: the natural quality of the resource in relation to standards (we call this measurement "background"), and the degree to which we maintain that quality or deviate from it (compliance).

Ambient air and water quality typically fluctuate, but we can expect their background ranges to remain within standard limits. Deviations from normal values alert us to a potential pollution threat somewhere in the system. Additional monitoring helps identify the source of the problem.

Environmental monitoring does something else, too. It tells us when we succeed and when we fail. By effectively monitoring our environment and heeding what we learn, we encourage good practices and identify bad ones. But monitoring alone yields only knowledge. It does not change anything.

ENFORCEMENT

Enforcement prompts us to toe the line and to correct mistakes. Without an effective enforcement program, the best permitting and monitoring systems will fail to protect the environment and our quality of life.

We learned this fact in the 1960's. Americans, in particular, became more aware of the dangers of unregulated development. But even though we began setting standards for air and water quality and establishing environmental bureaucracies, environmental improvement failed to show. Only when we put teeth into the laws did we begin to clean up polluting discharges and soften destructive practices.

To put it another way, we recognize the fallacy of reckless driving. But without the policeman to remind us of the law, we would drive much more recklessly than we do. So it is with our environment.

EDUCATION

The best designed program works only as far as people follow it. If we do not know the best route to take, we can hardly be expected to take it. Governments need to educate their citizens about their environmental needs and the best ways to meet those needs.

Governments do not always provide good information. Private organizations may fill the gap to some degree. But regardless of the way we learn, we need reliable information. Without it, many of us would stab blindly in the air. Others would not even care.

EXAMPLE

The most urgent need we have of our government is its example. We expect those who establish laws to guard them. This point may yield our greatest stumbling block to effective conservation. Our leaders do not lead by example.

History has recorded the Iraqi government's dumping of millions of gallons of crude oil into the Persian Gulf as a terrorist act. At least one writer has noted the conflict between the former Soviets' stated environmental awareness and their normal methods of treating industrial waste: landfills, sewers, and on site storage.[2]

Most of us understand chronically poor countries emphasizing economic development at the expense of environmental conservation. But we can hardly expect progressive leadership from governments in these countries.

Free industrialized nations, though, possess the training, capital, and technology to lead the world with an environmentally sensitive conscience. Too often, however, they choose otherwise.

Greed plagues governments as much as it plagues private individuals. Career politicians usually stress remaining in office over passing good legislation. For this reason, voters need to think for themselves, learn about the issues, and insist on responsible and accountable conduct by their representatives. Each of us needs to agree, "responsibility begins with me."

To do this, most of us need to alter our own conscience. We

need to look beyond our own interests and include a more universal perspective toward whatever issue we face. Many environmental issues must be addressed locally but may include impacts that reach beyond the locality. We cannot afford any longer just to talk a good game.

Perhaps 1990 points up the need for responsible government and citizen action toward the environment. That year appeared to be a quiet year for the environment, at least until the Iraqi crisis. Previous years' headlines of catastrophic oil spills and nuclear power accidents just did not appear. Spills occurred and droughts killed crops, but 1990 was passed over for a national disaster. Beneath the quiet surface, however, 1990 was not so benign.[3]

When we look more closely at the environmental gains and losses of 1990, we see that year dominated by double talk and deals. The United States Congress and the President promised an end to environmental indifference. Business and industry sang the praises of environmental responsibility. But as the year ended, many Americans realized the truth. Environmental jargon had been good business and good politics.

In June 1990, the United States Fish and Wildlife Service listed the northern spotted owl as a threatened species. This designation accomplished two things. It protected the owl by federal law. It also inflamed what may become the most intense economy/environment conflict in North America in recent years.

The full impact of this designation has yet to be felt in much of the United States. But a *Reader's Digest* article in 1992 credited this action already with costing the jobs of thousands of loggers and mill workers in the Pacific Northwest.[4]

There is even some legitimate doubt this owl needs special protection under the Endangered Species Act. Earlier, the Fish and Wildlife Service had determined it did not.[5] Then, environmental groups sued.

Naturally, hindsight is clearer than foresight. But perhaps it was at this point the federal government should have stepped in, not with favoritism, but with moderation. The real issue was how to conserve the resources (owl and forest) for their perpetuation and for their value to people.

To succeed, both sides would have needed to compromise. The government could have moderated that compromise and achieved a management plan protecting essential owl habitat as well as the livelihood of loggers and mill workers. But the government chose sides. The compromise still eludes us and the conflict still rages.

Other conflicts have raged for years over the use of America's national forests. These forests provide diverse recreational opportunities for millions of people. Endangered species, such as the South's red-cockaded woodpecker, find a home in our national forests. And national forests offer a bargain in wood to the timber industry.

In 1990, the United States Forest Service proposed to reduce logging in national forests and to emphasize wildlife and recreational values instead. In response, the timber industry pointed to Americans' pocket books. Thousands of families, the industry claimed, would be priced out of the housing market.

That claim is weak. Most of America's timber harvest comes from private forests and timber company land. Blaming the reduction of harvesting in national forests for a large rise in housing costs ignores the real culprit devastating our nation's forests: land development. Valuable forest land is being developed for subdivisions and business sites. And this resource reduction is permanent.

A force is at work in the United States that eventually may force most people out of the housing market. But that force is *not* the protection of national forests (more about that later).

One of the best barometers of our need for environmental leadership from our government can be seen in the United States' oil policies. Many of our environmental problems stem from our dependence on oil. During 1990, it seems, our demand rose while production declined. Nevada Senator Richard Bryan sponsored as one solution a forty percent increase in automobile gas mileage by the year 2001. This measure alone was predicted to save 2.8 million barrels of oil each day. But our federal government quashed the proposal.[6]

To his credit, President Bush declared a moratorium on offshore oil and gas drilling near the West Coast, the Florida coast, and a portion of the North Atlantic. This moratorium would delay oil exploration in these areas for a few years. Environmental groups anticipated a strong national energy policy encouraging conservation. Then came the Persian Gulf war. The year ended with no comprehensive energy policy.[7]

Purposely, the most desperate form of leadership we need from our government has been saved for last. Our government needs to provide an economic climate conducive to environmental conservation. Affluence encourages conservation; poverty discourages it.

No issue in America is more critical than our budget deficits. We ridicule "banana republics" for their astronomical inflation and indebtedness, placing them in the Third World since they must live off the resources of other countries. But the United States has become a debtor nation itself. And we seem to be satisfied with that status.

To see where we are headed economically and environmentally, we need only look at history and the world around us. History shows us our past mistakes, and our current situation shows us the fruit of those mistakes. We also can see what we did right.

Germany gives us a historical perspective to the current economic climate in the United States. Prior to World War I, Germany prospered under the strongest economy in Europe. Germans developed a flair for tapping international markets for a host of manufactured goods. By the beginning of World War I, Germany's industrial might surpassed even France's and Great Britain's.

The Treaty of Versailles in 1919 changed all that. This treaty required the German government to make economic reparations that were eventually fixed at $32 billion.[8] To accomplish this, Germany transferred most of its gold reserves to France and England.[9] Thus, German currency was stripped of any fixed support.

Hardship from chronic unemployment threatened uprisings in the cities, forcing the government to support the unemployed through transfer payments (entitlements and welfare). The German government was faced with supporting its population at home and making debt payments abroad at the same time.

Much of Germany's debt was held by foreigners, as is the case in the United States today. This gave the German government three tough choices: default (and lose trade with other nations), raise taxes (from a largely unemployed population), or inflate the currency (print money with nothing to back it up). They chose to print money.[10]

This decision eventually devastated Germany more completely than the war had. By October of 1923, the currency exchange rate had fallen to 4.2 trillion marks per dollar.[11] Germany's middle class was wiped out. We need to remember the middle class forms the backbone of constitutional government.

Hyperinflation finally destroyed the German government. The great depression of 1929 struck a confused and fearful Germany developing a rising tide of socialism. In January of 1933, Adolf Hitler was invited to take over the government. National Socialism had taken over.

Today we see Germany's experience being repeated over much of the earth. Argentina began borrowing heavily from foreign banks in the early 1970's. By 1980, Argentina owed $44 billion, a debt almost equal to its gross national product.[12] Interest payments alone required half of the government's income.[13]

By the mid-eighties, Argentina faced the same choices Germany had faced after World War I. As Germany did, Argentina chose to print more money. By 1990, Argentina's annual inflation rate had soared to 5,600%.[14] By then, savings were wiped out, the currency was worthless, and much of the population had lost everything.

By all rights, the United States should already have followed the way of Germany and Argentina. In *The Coming Economic Earthquake* (Chicago: Moody Press, 1991), Larry Burkett discusses enough economic facts about the 1990's United States to sober any thinking person. Among them:

1. The U.S. government's income is approximately $1.4 trillion per year.
2. The U.S. government spends approximately $1.8 trillion per year.
3. Our 1991 "on budget" and "off budget" federal debt totaled about $6.3 trillion (greater than our gross national product of about $4.3 trillion).
4. At current rates, federal deficits will add about $7 trillion in additional debt by the year 2000.
5. A national debt close to $20 trillion (about the year 2000) will require almost 200% of all personal income taxes to pay just the interest (figured at a 33% tax rate).
6. Consumer debt exceeded $794 billion in 1990.
7. There were 685,439 personal bankruptcies in 1990.
8. Business debt exceeded $700 billion in 1990 and required about eight percent of gross incomes for the interest alone.
9. A 1990 audit of the nation's banks determined 435 banks to be insolvent (the twenty largest represented almost $2 trillion in depositors' funds).
10. The National Organization of Life and Health Insurance Guaranty Associations declared 113 insurance companies to be insolvent or impaired between 1988 and 1991.
11. At an inflation rate of six percent, by the year 2000, the Social Security shortfall will be about $400 billion annually.
12. Eighty percent of all Americans receive some kind of government subsidy.
13. Twenty percent of all Americans work directly for the government (federal, state, and local).
14. U.S. currency is not backed by any fixed-value assets.
15. The American taxpayer works five months just to pay federal, state, and local taxes.

America's economic situation is frightening. The implications for our environment are just as frightening. Mr. Burkett points out the historical truth that freedoms are often forfeited during bad economic times.[15] We also know environmental quality suffers during such times.

Just recall what has already been said about the environmental situation in Brazil, Haiti, The Philippines, and the former Soviet Union. A bad economy thwarts a sensitive environmental conscience. Immediate human needs for food and housing outweigh environmental needs that may extend for decades. We can mourn this fact of life, but we cannot change it. Consequently, the single greatest sacrifice we can make for a healthy environment is the sacrifice we make for a healthy economy.

Environmental awareness is a luxury. We in the United States have enjoyed this luxury because a large portion of our population has been free from fear about the basic needs of life. Most of the rest of the world has not been so fortunate. We now stand to lose this fortune ourselves.

Already we hear cries to abandon special protection for endangered species because of the economic impact of such protection. Making matters worse, the Environmental Protection Agency seems to have elevated rules to an end in themselves rather than seeing that rules are means to an environment conducive to human health and well-being. Far from encouraging conservation, unreasonable federal strategies have forced a number of successful businesses to relocate outside the country.

Sometimes we need to come down hard on a polluter. But arbitrary heavy-handedness from EPA and other agencies feeds a backlash of anti-environmental sentiment. In this battle, we all lose—the economy, the environment, and the citizen.

Looking ahead, many of us see an environmental crisis dwarfing any we have experienced so far. We have improved air and water quality in the United States, though at terrific expense. We should not feel safe, however. Development pressures around the world are compounding. Vagaries of weather and climate add to human and environmental stresses. Natural disasters continue to occur. When these factors combine with an economic disaster, environmental disaster will surely follow.

A few years ago, Americans rejected the suggestion to sell many of the vast land holdings of the United States government.

We did not want national treasures eradicated. But when people get so deep in debt they cannot pay, they must either default or liquidate their assets and pay this to their creditors. A government is no different.

We can afford to sell some federal holdings, such as unnecessary consulates, obsolete military properties, underutilized office buildings, and useless government bookstores. These properties are not treasures, but national parks and forests are. If we do not control the federal budget and get out of debt, we will hear again the rumblings to liquidate many of our national assets. And next time, the audience will be more receptive.

A pessimist will look at our present situation and exclaim, "We're lost!" An optimist, however, will affirm, "Until the worst happens, we have hope." A healthy environmental conscience needs optimism. Bringing effectiveness out of that conscience requires action.

Chapter Nine explains steps each of us can take to shape a healthy environmental conscience. Our government needs to provide an economic climate where this conscience can bear fruit. This climate will require sacrifice—and not just by a few. Stewards will declare the benefits are worth the sacrifice. The following measures are offered as helpmates to generate a climate conducive to a healthy economy/environment marriage. Anything less probably will not succeed.

1. Congress must balance the budget.

This feat will require extensive cuts across the board in welfare, defense spending, federal payrolls, and subsidy programs—all of them. The Gramm-Rudman Act requires the government to live within its income and balance the federal budget. So far, Congress and the President have evaded these laws by shifting many items "off-budget," by overcalculating government income, and by robbing the Social Security Trust account to pay for budget deficits.[16]

Voters need to insist their representatives stop these practices. An even more effective measure may be a constitutional amendment requiring a balanced budget that includes all federal spending.

Congress also needs to give the president a line-item veto. As it stands now, the President cannot cut specific areas of overspending. He must reject the entire budget (an unlikely decision). A line-item veto will correct this costly foolishness.

2. Our government must get out of debt.

The answer does not lie in diverting nonbudget funds, such as Social Security, to the general budget. This just postpones pay day and risks a large segment of our population (the elderly). Nor does the answer lie in more borrowing, especially from foreign investors. Foreign investors already fund about twenty per cent of our annual deficit and own about $300 billion of our national debt.[17]

A limit exists even with foreign credit. Presently, our government competes with private industry for these funds. Eventually, the cost of this credit, due to supply and demand, will exceed the ability of business to pay. Then businesses will be forced to sell out to foreign investors to remain in business.[18]

The only strategy that will get us out of debt is the same strategy that will balance the budget: reduce the cost of government.

3. Our government must never again monetize its debt.

Printing money with no equity backing is so enticing only a tough federal law will prevent it. Voters need to insist on this law. Right now monetizing tempts Congress with an "out." Eliminating this possibility will give Congress and the President more incentive to balance the budget.

Germany and Argentina suffered the ravages of hyperinflation, which always follows monetizing a government's debt. Only when the printing of new money was forbidden, did Argentina even begin to regain control over inflation.[19]

4. Our government must provide tax reforms that favor business investment and personal saving.

This suggestion will raise some hackles. Americans always want to "gig" the rich. But at least two problems follow excessive taxing of wealth.

Only the wealthy have the funds to invest in America's economy. If the government takes this wealth away, it also takes away this country's investment capital.

In addition, excessive taxing prevents people from moving up the socioeconomic ladder. We Americans are so afraid someone is going to get rich, we eliminate even the possibility for most of us. Then we wonder why so many of us depend on the government for support.

As with rich individuals, we want to tax "greedy" businesses. We forget basic economics. No business pays taxes. Consumers pay the taxes. With additional funds from tax reforms, business and industry can invest in modern environmental controls and compete more favorably in the world's marketplace.

5. **We need to offer incentives to business and industry to improve their environmental controls.**

Presently, our government confines most of its incentives to punishment. We pay for this strategy daily through costly litigation, anti-environmental sentiment, and relocating outside the country. And we still pollute.

But human nature includes a quirk we can use. We perform better with rewards than with just punishment. Incentives, within reason, free up funds for companies to reinvest and to improve their operation. Without throwing away the rule book, we can encourage companies to practice better environmental management.

6. **We need to revise federal, state, and local permitting systems.**

Permits benefit people and the environment indirectly. They reduce our haste and cause us to consider alternative practices. But these benefits cost us dearly. A Business Roundtable study found some industries spent three years or more and as much as $300,000 to obtain permits for one plant.[20]

Such economic impacts can hurt rather than help the environment. Rather than build a new plant with modern pollution controls, a manufacturer may decide to maintain an outmoded plant. It may prove better, yet, to move out of the country. Either way, no environmental improvement has been realized, and jobs may have been lost. Together with tax reform, overhauling the permitting process will provide an economic impetus to improve environmental quality over the years.

FOLLOW THE LEADER

The Bible teaches, and science confirms, that where the head goes, the body follows. When our national government declines to make the tough choices required for a healthy environment, we can hardly expect states to do much better. And we need not expect the majority of Americans to lead private lives of environmental sensitivity.

Even so, the United States possesses an advantage over most other countries. We elect our leaders. We can address the lack of governmental leadership by electing genuine leaders and insisting they lead. This means the voter's conscience needs to change first.

The federal government alone cannot turn the tide of greed and indifference so prevalent in modern society. State and local governments can address some issues more effectively than the federal government can. Generally, though, we have failed on the state and local levels about as miserably as on the national level. We have depended on regulation, or even politics, rather than on conservation. Again, we have deferred the hard choices.

We need to strike a balance between the immediate needs of people and the need for long-term environmental conservation. We also need to learn how to live with a diminishing natural resource base. We cannot afford to be selfish. But selfishness is bound so tightly in human nature, we need a radical change of conscience to eradicate this blight. Nothing less will do.

Sometimes we need to say, "No!" We may have gotten away with an activity before. But now we see that, even though someone may profit from this activity, many more will suffer.

A public utility or industry, for example, can keep costs

down by only minimally treating its waste water before releasing it to the environment. The environment may absorb this pollution for years. Sooner or later, though, we experience the ill effects of a poor environmental conscience. Sludges blanket river bottoms. Algae choke lakes. Toxic compounds stunt or kill vegetation and threaten people. Some practices should be abandoned.

On the other hand, we cannot avoid environmental impacts altogether. Nor should we try. All living things alter their environment. If we choose to, we can devise a system of environmental management that will allow needed development while prohibiting truly destructive practices. This type of conservation system will allow our "yes" to mean yes and our "no" to mean no.

Since few of us willingly restrict our monetary endeavors, government needs to help us make this tough choice. It will not come easily. Come it must, however, if we are to conserve our environment for future generations.

But governing citizens is not enough. The United States government must address its budget deficits now and get out of debt. Otherwise, no amount of research, or permits, or monitoring, or enforcement, or public education will protect America's natural resources, including its human ones.

NOTES TO CHAPTER EIGHT

1. P. Aarne Vesilind and J. Jeffrey Peirce, *Environmental Pollution and Control*, 2nd ed. (Ann Arbor, Michigan: Ann Arbor Science Publishers, 1983), p. 141.
2. Dennis R, Poulsen, "The Soviet Environmental Dilemma," *Journal of Environmental Health* (Spring 1991): 31.
3. "The 23rd Environmental Quality Index," *National Wildlife* 29, No. 2 (Feb.-March 1991): 33.

4. Randy Fitzgerald, "The Great Spotted Owl War," *Reader's Digest* 141, No. 847 (Nov. 1992): 92.

5. Ibid.

6. "The 23rd Environmental Quality Index," *National Wildlife*, p. 38.

7. Ibid.

8. Bryce Lyon, Herbert H. Rowen, and Theodore S. Hamerow, *A History of the Western World* (Chicago: Rand McNally and Company, 1969), p. 716.

9. Larry Burkett, *The Coming Economic Earthquake* (Chicago: Moody Press, 1991), p. 71.

10. Ibid., p. 74.

11. Ibid., p. 75.

12. Ibid., p. 76.

13. Ibid.

14. Ibid., p. 78.

15. Ibid., p. 142.

16. Ibid., p. 118-119.

17. Ibid., p. 162.

18. Ibid., p. 163-164.

19. Ibid., p. 79.

20. Julie Sullivan, ed., *The American Environment* (New York: The H. W. Wilson Company, 1984), p. 209.

CHAPTER NINE

SHAPING OUR CONSCIENCE

"THE TIMBER INDUSTRY, ENVIRONMENTALISTS
AND THE GOVERNMENT HAVE JOINED FORCES TO
SAVE A MICHIGAN (KIRTLAND'S) WARBLER."

BRADENTON (FLA.) *HERALD*, 12 JUNE 1994.

Environmental responsibility is possible even in a world driven by financial lust.

If we accept that tenet, we establish a reference point from which we can shape our conscience to be more environmentally astute. Not that we simply follow so many steps to a sound conscience. Most of us need a change in attitude and conduct so complete our life-styles change.

Humans form an integral part of the earth's environment. We also can adapt that environment to suit our needs. Far from being a catastrophe, this ability can help us conserve rather than destroy. However, generations have been taught to value the environment only as a commodity. This needs to change.

Most of us probably wish we could wave a magic wand and miraculously clean up the air and water and reclaim contaminated soils. We hunt for mitigation measures that will successfully replace lost wetlands. We hope for some mysterious key to elevate endangered plant and animal populations above their threatened levels.

Except for a pitifully few success stories, we have wished and hoped in vain. Even these successes have been achieved more through hard work and sacrifice than through a magic formula. Sometimes nature's resilience alone has staved off disaster.

We have hunted for environmental salvation in all the wrong places. Science can increase our knowledge about the environment and identify techniques to conserve it. Technology can produce the tools we need to achieve our goals. Economic management can pay for the work. But without a change of heart, we will not commit to employ our generations of economic, scientific, and technological progress to conserve our environment for the future. We will keep on living for today.

A CHANGE OF HEART

Some readers may scorn this section as too religious, too emotional, too unscientific, too whatever. Yet, please keep in mind the radical change we all need. Such a change requires a radical force, a power beyond our mind and muscles. We cannot simply drop environmental proverbs and quote ecological principles. We need to act with wisdom and determination and, sometimes, with deference. Then we need to encourage others to act wisely, too.

Nowhere else do we find this perspective but in the concept of stewardship. We work to achieve our needs and dreams, to be sure, but we are "minding the store" for someone else, too.

We will never develop a steward's heart until we look beyond ourselves. Stewardship demands trust, and trust requires someone to trust. Generally, this kind of trust is directed toward God. Not everyone accepts the reality of an omniscient and omnipotent God, but even this belief can change. Belief involves choice. Too often, people choose not to believe in God because they think they cannot see any evidence of God in the world around them. They are mistaken.

Perhaps the Apostle Paul said it best: "For since the creation of the world His invisible attributes, His eternal power and divine nature, have been clearly seen, being understood through what has been made, so that they are without excuse" (Rom. 1:20, NASB). They just have not recognized what is evident.

Without accountability to a higher authority, we give ourselves little incentive to protect the environment for future generations. Virtually every human being wrestles with greed. Some just give in more easily than others. Some of us blatantly scorn our children and grandchildren by a grossly selfish life-style. Others, though not so openly selfish, still live for the moment. We all have demonstrated we do not have in ourselves the strength to overpower this predisposition. We need power from a better source.

The simplicity by which we gain God's power in our lives stops many people cold. They cannot believe it is so easy. But it is. We receive God's presence and power by asking. Then we believe it, "for the Scripture says, 'Whoever believes in Him will not be disappointed' " (Rom. 10:11, NASB). It is that simple.

This book was not written to be a religious text. Environmental problems cripple us all, regardless of religious preference, political affiliation, or anything else. But no matter how hard we try, we cannot escape the need for grace, strength, and wisdom beyond what we can conjure up ourselves if we are to solve more than the most elementary environmental problems. Human history shows we create more pollution even as we clean up some. We need to break that cycle. Here is how we can do that.

BE SATISFIED WITH LESS

For most of us, this might be the most radical of all ideas. We work for raises. We advertise for business. We define success with dollars. No matter how much we have, we want more. Then we look for excuses for our troubles.

Many people are acutely and chronically poor. We call them lazy. Millions of dollars in resources are tied up by defense contractors. We call them warmongers. Air and water are polluted by industries. We call them gluttons. Our world loses natural resources. We call it overpopulated.

There may be a grain of truth in those judgments at times. The root of our environmental problems, however, comes from none of those situations. We excessively alter our environment because we overconsume. Resource protection does not stand up well in the face of an insatiable market.

The oil and gas industry provides a good object lesson.

Today, it is fashionable to pick on the oil companies. We call them greedy, insensitive, and worse. These companies probably earned much of this reputation and, certainly, have contributed a huge share to the world's pollution. As time goes on, we regulate oil companies more tightly. Still, we fall short of requiring of them genuine accountability in many cases.

As consumers, we have decided a cheap energy source is worth the pollution risk when alternatives might cost us more at the pump. A few regulations will not offset the risks from a ravenous petroleum market. But we are loathe to use less oil.

Some public utilities have stumbled onto the idea that raising the price of potable water will reduce consumption without reducing their income. Perhaps, it even works to a degree. But a long-term solution to water consumption woes calls for more than periodic rate hikes. That is the easy way. That solution also hits some people very hard while hardly touching others. Some folks can afford to pay any price the utility can afford to charge without reducing consumption. Others with lesser means may have to do without something.

As ambient water quality declines and demand rises, the days of cheap water will end. Unlimited potable water may become unavailable at any price. Do not scoff. Rationing of one kind or another has been practiced around the world. Even in the most affluent country in the world, the United States, water restrictions have become a way of life in some areas. Rationing may be just around the corner.

If water rationing comes upon us, it will declare the lack of resolve of our citizens to protect this most universal and indispensable resource. Rationing does allocate water more fairly than pricing some people almost out of the market. But rationing declares failure, as well.

Whether oil, wood, water, or any other natural resource, if we use less, we will help perpetuate the resource and reduce pollution from its utilization. Choosing to be satisfied with less will not hurt anyone. If we all choose this strategy, we will help everyone.

CREATIVE THINKING

Too often, we employ the same old strategies we have tried before. We expect certain industries to pollute the air, so we clean, or "scrub," emissions with processes so intricate they would baffle Rube Goldberg. We pay a bundle for them, too. We do this even though we know pollution control is more effective the farther we go up the process line. Apparently, we see emissions, so we attack emissions. But we still pollute the air.

We may never learn to build an industrial plant that needs no emissions control. We can, however, minimize the production of emissions, thus making emissions control more efficient. Some creative people have already made progress in this area. The rest of us need to encourage such efforts.

When faced with an environmental problem, wisdom finds a practical solution that works. Various habitats disappear around the world each day. Government offers some measure of protection on public lands, but much of the world's unique habitats exist on privately owned land. Private interests generally must show a profit to survive. Private land can be mined, subdivided, paved over, turned into landfills, and developed in myriad other ways. But private land also can be bought.

Someone deduced a fair method of resource protection would occur through the purchase and holding of various tracts of unique habitat. The property owner would benefit from the sale, and many others would benefit from the resource protection. Private organizations, such as the Nature Conservancy, thus acquire unique habitat to preserve and perpetuate. And they do it without raising taxes.

The 500-square-mile Gray Ranch in southwestern New Mexico perpetuates a diverse ecosystem of desert grassland, mountain forests, floodplain woodlands, and a 44,00 acre prairie of blue gramma and buffalograss.[1] Within this preserve is reported a lengthy list of rare species: bald eagle, elegant trogon, peregrine falcon, ridge-nosed rattlesnake, Wright's fishhook cactus, and many more.

This land could have been trodden under the feet of cattle—a legitimate use. But a complex array of ecological communities would have been stressed perhaps to the point of breaking down. Thankfully, people recognized the economy/environment conflict.

A practical solution was found. The property owner was compensated through the purchase price, and a unique American resource was protected for generations to come. That is not a bad solution.

There is a potential negative side to this solution, however. Carried to an extreme, we could conceivably remove most undeveloped private land from marketable production. By doing so, we could eventually devastate our economy with our environmental zeal. Economics, itself, probably would prevent such large-scale purchase of land by conservation groups, but it gives us food for thought. Careful planning and judicious practice must be employed for this strategy to work.

We cannot afford to remove most of our land from economic production. Neither can we afford to eliminate most of the naturally functioning ecosystems from our planet. We need balance. Environmental strategies like the Gray Ranch have their place. Further abroad, the International Whaling Commission imposed a moratorium on killing minke whales. Apparently, some Norwegians continued killing them, only now for "research." In 1990, the United States Secretary of Commerce cited Norway under the Pelly Amendment to the Marine Mammal Protection Act and recommended an embargo on Norwegian fish imports.[2]

Across the ocean, sealers in Nova Scotia and Prince Edward Island have for many years harvested juvenile harp seals for their white-coated pelts. Tour groups now have begun to visit these same areas of Canada to view the same seals. These groups were reported to have generated about $600,000 in the region's economy, or about twenty per cent more than the value of the pelts.[3]

This specialized tourist industry and the suggested embargo turn on the same theory. Economics may encourage exploitation, so use economics to encourage conservation. With one strategy, a market was reduced. With the other, a new market was opened, a market not dependent on consumption. Strategies of both kinds have their place.

Creativity need not always be expressed in the negative, such as a regulation or prohibition. Creative thinking really shines when directed toward a positive environmental stance. Such a stance, though, can be hard to maintain. We may identify a problem easily enough. But overcoming the problem can tax the smartest of us and may very well come from the simplest idea.

For instance, the Philippines supply the majority of tropical marine fish to the aquarium hobby. Fishermen have stunned many of these fish with clouds of sodium cyanide gas squirted around a coral reef. Many fish never recover, and cyanide is poisoning portions of the Philippines' magnificent reefs. Cyanide also has been used to catch food fish in these waters.

The Philippines government has taken steps to curtail the

use of cyanide for fishing, though the fear of arrest may not deter many fishermen. They fish to survive in this chronically poor nation. But someone thought of a solution many Filipino fishermen might have considered radical: using a combination of dip nets and fence nets to capture reef fish.[4]

Filipinos were no longer just threatened with arrest but instead were offered an effective alternative method of fishing. Experienced netters then train others. Perhaps this trend in the Philippines will continue and spread elsewhere. But someone had to lead in addressing the problem.

Perhaps the most effective thinking comes only after we accept the fact that no government can resolve the plethora of today's environmental problems. Some problems, though touching many lives, are too small to warrant much government attention. Even with more significant problems, governments may be unable to pay for a public remedy. In this situation, individual initiative and creativity offer the only practical hope.

In 1972, Tulsa, Oklahoma, was losing its stately elm trees to city progress. Many of these trees had shaded Tulsa since the turn of the century. A Tulsa teenager took his objections to the city and persuaded a street commissioner of the importance of conserving an urban environment. But Tulsa, at that time, could not afford to replace trees removed for city street projects. So, citizens and businesses banded together to sponsor sites and beautify freeways with native trees and wildflowers.[5]

The city government cooperated. But private interests led the efforts to bring about environmental improvements in their own part of the world. We should commend them.

Far from being powerless, individuals hold the key to creative conservation. Each of us can act in an environmentally sensitive manner. We also can join others in directing our government and in carrying out positive environmental efforts right where we live. The most effective conservation begins at home.

CAUTIOUS LEARNING

Even the most creative thought can fail if it is founded on poor information or misunderstanding. Before we allege environmental mismanagement, we need to investigate to see if a genuine problem exists. Otherwise, we can look like fools and cause more harm than good. At the very least, we will waste precious time and resources for little benefit. We see this happening today with a certain element of solid waste.

Plastic, especially fast food packaging, bears the brunt of a misinformed public information campaign. To be sure, plastic litter is unsightly and poses a threat to certain species of wildlife. But plastic is not filling landfills; paper is. Overemphasizing the plastic component of solid waste has diverted our attention from the mountains of newspapers and telephone books stuffing our landfills.

Plastic can be energy-wise, too. In 1992, *Reader's Digest* reported a conclusion of California Futures, an economic and environmental research organization, about plastic containers. They noted, of nonrecycled containers, plastic takes less energy to manufacture than either glass or aluminum.[6]

The solution to a knowledge deficiency is to study before we speak. A wealth of information exists regarding just about every environmental problem we know of and some we only imagine. Yet, some people gather all their information from the most unreliable sources: the popular press and television.

Judicious use of these sources can help us. Some of the information in this book came from popular magazines and newspapers. But we need to remember the success of these enterprises does not depend on objectivity but on emotional appeal. The environment in which we live and raise families is too precious to entrust to our emotions. Workable solutions to complex problems require solid information.

Public libraries offer a convenient and inexpensive place to start. Larger libraries not only carry a host of books, but also many cultural, scientific, and historical periodicals. The United

States government floods the nation with publications, many of which address environmental issues. And we need not overlook people we know. Chances are most people know someone who is a true expert on some element of the environment.

Caution, though, is the watchword of learning. In our quest for environmental literacy, we need to remember each book and magazine we read and each person we talk to comes from a certain bias. Some "sources" display such obvious unreliability we can reject them out-of-hand. But most sources seem more convincing, and we need to chew things over before we swallow.

Consider government workers. Contrary to popular myth, most government employees are not double-talking, self-serving imbeciles. They are no better or worst than anyone else. Many are true experts. If we understand these people represent one side of an issue, usually research or regulation, we can glean a lot of useful knowledge from them. We pay our government to study. We may as well take advantage of what is learned.

Environmentalists approach matters from a different perspective. These folks are generally more protective, sometimes to an extreme. But we can learn a lot about human interaction with the environment from environmentalists.

Business and industry approach environmental issues primarily from a utilization perspective. As environmental consumers ourselves, we should be able to understand this perspective even if we do not appreciate the excesses it sometimes breeds.

If we examine the issues, understand the biases, and sort through the information, we can devise solutions to many problems we face every day. For this reason, the premise for this book was stated early in Chapter One. By recognizing that premise, the reader may more easily understand the common thread woven through this book and more easily judge its contents.

Study means work. To default in this responsibility means, at best, we will hover on the edge of effectiveness. At worst, we will blindly wreak havoc where a solution could have been found.

BECOME INVOLVED

It seems silly to think anyone would take the time to study an issue without wanting to implement what was learned. Sometimes, though, the best informed of us drop the ball. We have careers, families, schooling; and involvement takes precious time. But none of us need be an activist to become involved. In fact, too much high profile activity can muddy the water. But there are ways to help.

Conservation organizations run on money. Contributions, regardless of the amount, help identify and resolve some environmental problems. It is that simple. Of course, using this approach alone gives us the easy way out.

Personal support, as well as campaign contributions, keeps politicians in office. Personal support also publicizes environmental issues and encourages people to work toward a solution.

On September 23, 1989, 2,800 volunteers cleaned nine tons of debris from 176 miles of Maine's shoreline.[7] This record support, including children, generated a teacher's guide to water quality in the Gulf of Maine and donations of over $7,000 worth of goods and services from several businesses.[8] Personal support makes a difference even on one day.

We need to use caution in choosing which activities to support. Some environmental activism is so radical as to be counterproductive, at best, and sometimes terroristic.

An ugly rumor surfaced along Florida's west coast a few years ago. A group of "environmentalists," it was said, decided they would end the abuses of commercial netting of fish by locating the nets and sinking them. Apparently, those activists thought the fishermen would lose their expensive nets and stop fishing.

If this rumor was true, those activists do not know fishermen. They also ignored a horrifying truth. Sunken nets would continue to trap thousands of fish, crabs, and lobsters for years, and none of those creatures would be harvested for consumption. They would lie on the bottom of the Gulf of Mexico and rot. What sort of conservation is this?

Radicalism aside, one of the greatest hindrances to resource conservation is the failure to disseminate reliable information. The information is out there. Most people just never see it.

Volunteers can collate, package, and mail literature related to local environmental concerns. Conservation organizations can make sure newspaper editors get the facts about an issue facing the community. These editors still may prefer to rely on the emotional neighbor and downtrodden victim. But eventually, truth, when presented truthfully, has a way of winning out.

Businesses, too, can join the effort to conserve the environment. It is their environment, too. In 1991, encouraging news about the Dow Chemical plant in Joliet, Illinois, was reported. It was noted 75 acres of Dow's 840-acre tract was devoted to manufacturing. The rest was managed for the natural resources it contained.[9]

Conservation pays off. It also costs. Money can buy land, fund research, and publish educational material. But money does nothing of its own. Only when people act responsibly can progress be made in solving environmental problems.

We each have a role to play. Whether we act or not is our choice.

NOTES TO CHAPTER NINE

1. Bennett A. Brown, "Apacheria Revisited," *Nature Conservancy* 40, No. 5 (Sept.-Oct. 1990): 21.
2. "The Lineup," *Tropical Fish Hobbyist* 39, No. 8 (April 1991): 77.
3. "The Lineup," *Tropical Fish Hobbyist* 39, No. 9 (May 1991): 138.

4. Don E. McAllister, "A Glowing Future for Marine Aquarium Fishes," *Tropical Fish Hobbyist* 36, No. 12 (August 1988): 85.

5. Edwin Dobb, "Catch The Spirit," *Reader's Digest* 138, No. 829 (May 1991): 94-95.

6. Lynn Scarlett, "Don't Buy These Environmental Myths," *Reader's Digest* 140, No. 841 (May 1992): 102.

7. Kathryn J. O'Hara, *Cleaning North America's Beaches: 1989 Beach Cleanup Results*, ed. Rose Bierce (Washington D.C.: Center for Marine Conservation, 1990): 121.

8. Ibid.

9. "Rare and Well Done," *Tropical Fish Hobbyist* 39, No. 10 (June 1991): 133.

CHAPTER TEN

✦

THE THREE GIANTS

We all want to live in a clean and healthy environment. But how? We work so hard to kill, to waste, to pollute, and otherwise to alter our environment so as to make much of it inhospitable. But there is a way to expand our individual conscience to a local, national, and global level. Sound impossible? Committed men, women, and children achieve what seems impossible every day.

Modern societies employ three giant channels in a great effort to persuade people and direct their energies. These channels—education, technology, and economics—help us achieve all manner of outcomes, some good and some bad. People sensitive to environmental needs can employ these channels to persuade others and direct their environmental efforts around the world. What succeeds at home can then grow into a broader success.

The seed of freedom planted in Eastern Europe lay dormant for a long season. In time, though, the seed grew and blossomed and astounded the world with its fruit of victory over communist oppression. Our environmental conscience can bear similar fruit. But first we need to join together and learn to employ these three giants.

EDUCATION

The world does not lack for information. Almost every environmental issue we know about has been studied to some degree, and literature has been published on the matter. Most of us just never see it. We have failed to disseminate the very information we need to form objective opinions and develop practical strategies to resolve environmental problems. In some cases, we do not even know enough to prioritize the problems.

We desperately need to collect and disseminate helpful information. For instance, in Florida, new developments must employ stormwater management systems to collect surface runoff from rain events. Hence, subdivisions often are built around one or more artificial detention ponds. This system benefits the environment, and people, by slowing runoff and by improving the quality of the water before it leaves the site. But a frustrating marketing scenario has developed.

Developers sell property adjacent to these stormwater ponds as waterfront lots. Many buyers pay the higher prices without understanding what sort of water these lots front. Once a predetermined number of lots are sold, the developer turns over the project's maintenance to a homeowner's association. As more homes are built, more nutrient-laden runoff enters the ponds. Algae growth eventually explodes into great blooms of green water, sometimes depleting the oxygen supply and killing fish.

Eventually, bewildered residents call the local pollution control agencies for help. At this point, there is little to do except wait for the algae bloom to subside and the system to return to normal—until next time.

The facilities represented in this scenario did exactly what they were designed to do. They collected and detained stormwater together with the nutrients and pollutants associated with urban runoff. The problem was that the homeowners had not known what kind of ponds they had bought and what could be expected to occur in them. And they were not happy with the results.

One pollution control agency pulled together information from various sources and produced a pamphlet that explained the function of stormwater facilities and what could be expected of them. The pamphlet was not fancy. But in it, tips were given about managing these facilities to reduce the unpopular aspects of stormwater collection. Rather than wait for people to learn of the pamphlet, this agency mailed them to local homeowner's associations and to other groups to pass along to their clients.

Stormwater complaints did not cease altogether. However, many residents felt there was something they could do now to manage a water body they hoped to be an amenity to their homes.

Specific applications, such as this one, usually achieve specific results. Environmental education also accomplishes less tangible result. It raises social environmental awareness. Before we can carry out sound environmental policy on a large scale, we must teach people about the need for such policy.

None of us escapes the need for education. But Third World citizens, in particular, need to know that too-rapid economic growth can spawn dire economic and environmental repercussions in the future. They will learn this truth sooner or later. Perhaps they can learn from the mistakes the rest of us made and avoid some of the environmental problems industrialized nations have raced to create. Education will be the key.

In 1985, a cooperative effort began between The Pragma Corporation, the government of Belize, and the United States Agency for International Development to "Increase the Productivity Through Better Health (IPTBH) of Belizeans."[1] This effort included a component to insure adequate safe water and sanitation to a large segment of rural Belize. But first the participants had to educate those they came to help.

Strong community involvement grew from public meetings, family gatherings, and other activities. This type of system can work. Villagers were reported to have continued to show interest in health care and to have expanded that interest to include nutrition, dental hygiene, and related health topics.[2]

The worldwide need for environmental education begs for truth, not indoctrination. To answer that need, we must integrate the various aspects of environmental health into a balanced strategy to protect our world and improve our quality of life.

Health, sanitation, and resource conservation go hand in hand. To apply such principles, we need to integrate ethics, history, social science, and natural and physical sciences into a program adaptable to diverse socioeconomic situations. We share a common heritage and a common environment. Our response to this fact will determine the quality of our environmental policies.

TECHNOLOGY

We will not solve modern environmental problems without employing modern technology. Needs are too great and problems too complex. Gone are the days when we could assume "dilution is the solution to pollution."

We still rely heavily on cleaning up polluting discharges, but this strategy has failed to keep up with the pollution we generate. We need a new route to pollution control. Many people believe the answer lies in technology that promotes economic growth without environmental degradation.

Perhaps this solution sounds improbable. For generations, we have witnessed economic development degrade the environment. People live in slums, drink tainted water, and breathe polluted air. Some people even pay for progress with their lives. We need to remember, however, altering the environment does not always degrade the environment.

The world has changed tremendously during the course of

human history. Natural events have spawned much of this change. People have changed things even more.

Not all of this change has been bad. Modern farming methods, for instance, protect soil and water much better today than in years past. Much of our environmental failure seems to have resulted from assuming we must pollute to grow.

Prompted by legal restraints, technology has greatly reduced polluting emissions from automobiles. We need to applaud this progress. We also need to develop alternatives that produce less congestion and fewer emissions in need of cleaning. Only then will we have taken real steps toward environmentally sensitive transportation. Chances are, no single measure will solve the problem. A multifaceted problem usually requires a multifaceted response.

Natural resources are not evenly distributed around the world. Neither is technology. Many people in industrialized countries believe wealthy nations need to share skills and resources with poorer nations to resolve environmental problems affecting the whole world. We need to share technology, as well.

Yet, political and economic instability and massive debt bar many Third World nations from the technology market. Informed and creative thinkers can help structure a program to share peaceful technology wherever it is needed in the world without wrecking the supplying nation's economy.

Just as terrorism is impossible to combat without international cooperation, worldwide environmental problems cry out for similar cooperation. Otherwise, some countries will grab the meat, and others will fight for the bones.

A modern application of an ancient fishing method, netting, offers hope to preserve the integrity and economic importance of the Philippines' coral reefs. To work, this technology had to be shared.

ECONOMICS

We can devise the most efficient antipollution technology imaginable. We can teach people how to use it. But this technology will not improve one molecule of air or water unless we employ it. Applying modern technology takes money, sometimes a lot of money. Thus, some people judge environmental responsibility as not being worth the cost. Many of us, however, disagree.

Far from being the cause of environmental problems, capitalism offers the only means of paying for their remedy. The State controlled Soviet economy for decades. Yet, the former Soviet Union was plagued with some of the worst environmental crises in the world. Most Third World nations face more basic needs, such as safe drinking water and sanitation facilities. And most Third World nations cannot pay for much improvement.

Several environmental organizations have joined with a few Third World countries in debt-for-nature swaps. These swaps aim at reducing both tropical deforestation and Third World debt, though success has been minimal.[3] The funding of these organizations is limited, and it is unlikely any nation will trade off all its rain forests or other natural resources. And it should not. These swaps could be carried to an extreme and become more of a problem than a solution. But they can be useful within reason.

Sooner or later, though, the industrialized world will almost surely have to share a greater degree of economic and technological assistance with the Third World (*assistance*—not handouts). Large scale environmental problems hurt rich and poor alike.

The key word is share. Many citizens of wealthy nations resent aiding poorer nations because they perceive poor "foreigners" to be lazy and ungrateful. Many Third World citizens resent aid they perceive to come from imperialist countries.

Most countries desire autonomy, and rightly so. But some

environmental problems cross international boundaries. Such problems demand international solutions.

This is not to suggest the world should join in one government or one economy. That fantasy is doomed already. But countries such as the United States and Canada can initiate agreements to resolve their parts in environmental abuses affecting one another.

A basic strategy to reverse the economic decay in the United States was outlined in Chapter Eight. Before we can have much positive impact elsewhere, we need to clean up our own house. Success at home can then be spread abroad. We can balance the economy and the environment. Whether we do or not is a matter of choice.

Ultimately, the love of money yields all manner of evil. Yet, money is only a tool. Economics can he used to help people grow. Economics also can be manipulated to prevent poor people from rising out of their poverty.

We witness daily the failure of massive foreign aid programs to make any real difference in the lives of the world's poor. The money is going somewhere, but not to those who need it most. Neither local economies nor the environment improve from such foolish spending. However, $100 in the right hands can change dreams to reality.

In Ubate, Colombia, the Trickle Up Program, a private nonprofit organization, provided the Perdomo family with $100 and a little advice. The Perdomo's then started their own hair roller factory. Neighbors were so impressed, it was noted, they started five other businesses.[4]

Economic ripples spread from such endeavors. Successful families can afford more and better food purchased from their neighbors. They can afford better housing built by neighborhood labor. And most such endeavors are more environmentally sensitive than such Third World jobs as trapping rare birds for the pet trade.

To balance the economy and the environment, we must consider both to be equally valuable. We may wish the World Bank had paid more attention to environmental impacts when funding economic development projects in the past. But we cannot change history. We can, however, plan for the future.

Sooner or later, Third World governments and citizens will know great economic development cannot be sustained once crucial natural resources are depleted. But many people, even in industrial nations, have not learned the economy/environment connection.

We all need to understand this connection. We can employ education, technology, and economics to raise our environmental consciousness and bring about more responsible behavior. The chance for failure is great. The need for success is far greater.

Environmental revival lies mostly in the heart of the individual. We can shape our conscience to overcome the root of infection that has led to our part in environmental (and human) abuses. Then we can encourage others to do the same. The choice is ours to make.

NOTES TO CHAPTER TEN

1. Terri Jenkins-McLean, "Lessons Learned From a Third World Water and Sanitation Project," *Journal of Environmental Health* (Spring 1991): 34.
2. Ibid., p. 38.
3. Lewis Mennerick and Mehrangiz Najafizadeh, "Third World Environmental Health," *Journal of Environmental Health* (Spring 1991): 26-27.
4. Carolyn Males, "$100 Dreams," *Reader's Digest* 138, No. 829 (May 1991): 113.

CHAPTER ELEVEN

✦

A PERSONAL NOTE

Thank you for sticking with this book to this point. I trust you have been challenged rather than dismayed. Environmental problems around the world are real, and remedies in many cases are expensive, perhaps in a few cases non-existent.

We still have plenty of room for optimism, however. We have experienced some successes and have learned more about our environment than we have ever known before. We have the knowledge, the resources, and the technology to live lifestyles of environmental integrity. We seem to lack only the will.

Our conscience determines how we live. We do not live for ourselves alone—at least, we should not. But, for some reason, we have not trusted ourselves to make the tough choices our life-styles have left us with.

I think, perhaps, most of us are still too self-focused. I know I am. Daily, I struggle with the same desire: I want what I want when I want it. Still, I know deep down this desire is neither reasonable nor good for me. I am learning to wait.

Probably that is the only way we will reign in our overconsumption. Waiting means patience. But without patience, we can devastate the environment as well as the economy. Patience breeds optimism, however, and optimism fuels cooperation.

In September of 1992, I watched on television a report about some self-proclaimed "conservationists" aboard a ship named the "Sea Shepherd." This ship's crew purposely rammed Japanese fishing boats and sank some of the nets from those boats. The "Sea Shepherd" then returned to port, and the crew bragged of their exploits to the media.

Without a doubt, some Japanese fishermen kill sea birds and dolphins and overharvest certain species of fishes. But the Japanese are not alone in this abuse. Neither is this a problem only with commercial fishermen.

The fallout from such aggression can cripple rational environmental programs. Morally, the crew of the "Sea Shepherd" was wrong. I can understand their frustration, but their violence was unjustified and accomplished no lasting benefits. It may even invite retribution.

This crew did accomplish environmental abuse as bad as that of the fishermen they attacked. The nets they sank will continue to catch fish and other sea creatures for years to come. Only, these animals will not be harvested for human consumption. They will remain in the nets and rot while acting as bait to lure more animals into the nets.

What kind of conservation measure is this? Where does this behavior offer any hope for cooperation? Terrorism will never solve environmental problems. It will only feed anti-environmental sentiment and legislation.

Fiscal irresponsibility also will block environmental conservation. Please, remember a healthy economy promotes a healthy environment. A poor economy discourages an environmentally sensitive conscience.

I am not old enough to remember the Great Depression that began in 1929. You may be. You may remember how an unemployment rate of forty percent almost wiped out the middle class in the United States. A "New Deal" was tried for almost a decade. It failed. The Great Depression did not end until the United States entered World War II.

Healing the American economy today will take personal and governmental sacrifice across the board. Surely, though, this sacrifice is preferable to the sacrifices of war. Ultimately, this sacrifice will yield an economic environment conducive to improving the natural environment we all must inhabit.

Please keep in mind, our actions define our character. A sensitive environmental conscience, a steward's conscience, can be achieved. I believe you now know how to make such a conscience a reality in your life. Please, do not give up.

May God bless your efforts in achieving a life-style of environmental integrity.

Wishing you the greatest success,

Gary Cochran

Everyone makes some decision about God. The following Scriptures proclaim the simplicity by which we receive the personal change of heart we all need. Reading these verses will not hurt. Embracing them will change your life. (*New American Standard Bible*, The Lockman Foundation).

1. "There is none righteous, not even one" (Rom. 3:10).
2. "For all have sinned and fall short of the glory of God" (Rom. 3:23).
3. "But God demonstrates His own love toward us, in that while we were yet sinners, Christ died for us" (Rom. 5:8).
4. "For the wages of sin is death, but the free gift of God is eternal life in Christ Jesus our Lord." (Rom. 6:23).
5. "That if you confess with your mouth Jesus as Lord, and believe in your heart that God raised Him from the dead, you shall be saved; for with the heart man believes, resulting in righteousness, and with the mouth he confesses, resulting in salvation" (Rom. 10:9-10)

BIBLIOGRAPHY

1. Ainsworth, Susan. "A Current Affair." *Resources* 12 (1990): 3-4.
2. Beever, James W. 1988. *The Effects of Fringe Mangrove Trimming for View In the South West Florida Aquatic Preserves.* Report for Florida Department of Natural Resources.
3. *Bradenton* (Fla.) *Herald*, 20 May, 1991.
4. Brown, Bennett A. "Apacheria Revisited." *Nature Conservancy* 40 (1990): 16-23.
5. Bryan, Hal D. "A Saltwater Wetland In Northeastern Kentucky." In *Proceedings of The Fifteenth Annual Conference on Wetlands Restoration and Creation*, edited by Frederick J. Webb, Jr. Tampa, Florida: Hillsborough Community College,1988.
6. Burkett, Larry. *The Coming Economic Earthquake.* Chicago: Moody Press, 1991.
7. Cobb, Charles E., Jr. "Living With Radiation." *National Geographic* 175 (1989): 403-437.
8. Culliton, Thomas J., et al. *50 Years of Population Change Along the Nation's Coasts 1960-2010.* Rockville, Maryland: U.S. Department of Commerce, 1990.
9. Dobb, Edwin. "Catch the Spirit!" *Reader's Digest* 138 (1991): 94-98.
10. Eisenbud, Merril. *Environment, Technology, and Health: Human Ecology In Historical Perspective.* New York: New York University Press, 1978.
11. "Environmental Outlook in the USSR." Newsfront. *Resources* 12 (1990): 2.
12. Fernald, Edward A., and Patton, Donald J., eds. *Water Resources Atlas of Florida.* Tallahassee: Florida State University, 1984.

13. Gafafer, W.M., ed. 1966. *Occupational Diseases: A Guide To Their Recognition.* U.S. Department of Health, Education and Welfare, Public Health Service Publication no. 1097. Washington, D.C.

14. Gist, Ginger L. "The New 'Dead Sea': Persian Gulf Oil Spill Advances Ecological Clock." *Journal of Environmental Health,* Spring 1991, pp. 20-22.

15. Jenkins-McLean, Terri. "Lessons Learned From a Third World Water and Sanitation Project." *Journal of Environmental Health,* Spring 1991, pp. 34-38.

16. Knap, Alyson Hart. *Wild Harvest: An Outdoorsman's Guide to Edible Wild Plants In North America.* Toronto: Pagurian Press Limited, 1975.

17. "Long Island Sound Study Proposes Interim Actions to Combat Hypoxia." *Coastlines* 1 (1990-1991): 4-7.

18. Lyon, Bruce; Rowen, Herbert H.; Hamerow, Theodore S. *A History of the Western World.* Chicago: Rand McNally and Co., 1969.

19. Maehr, David S. and Belden, Chris. "The Endangered Florida Panther." *Florida Wildlife* 45 (1991): 15-18.

20. Males, Carolyn. "$100 Dreams." *Reader's Digest* 138 (1991): 111-114.

21. McAllister, Don E. "A Glowing Future for Marine Aquarium Fishes." *Tropical Fish Hobbyist* 36 (1988): 84-86.

22. Mennerick, Lewis and Najafizadeh, M. "Third World Environmental Health." *Journal of Environmental Health,* Spring 1991, pp. 24-29.

23. Mitra, Sachinath. *Mercury In The Ecosystem: Its Dispersion and Pollution Today.* Switzerland: Trans Tech Publications Ltd., 1986.

24. Murphy, Robert. *Wild Sanctuaries: Our National Wildlife Refuges—A Heritage Recorded.* New York: E. P. Dutton and Co., 1968.

25. O'Hara, Kathryn J. and Younger, Lisa K. *Cleaning North America's Beaches: 1989 Beach Cleanup Results.* Edited by Rose Bierce. Washington: Center for Marine Conservation, 1990.

26. Poulsen, Dennis R. "The Soviet Environmental Dilemma." *Journal of Environmental Health*, Spring 1991, pp. 31-32.
27. Plage, Dieter and Mary. "Return of Java's Wildlife." *National Geographic* 167 (1985): 750-771.
28. Rathje, William L. "Rubbish." *The Atlantic* 264 (1989): 99-109.
29. "Rare and Well Done." *Tropical Fish Hobbyist* 39 (1991): 132-135.
30. Scarlett, Lynn. "Don't Buy These Environmental Myths." *Reader's Digest* 140 (1992): 100-102.
31. Sullivan, Julie, ed. *The American Environment.* New York: The H. W. Wilson Co., 1984.
32. Tampa Bay Regional Planning Council. 1986. *Habitat Restoration Study for the Tampa Bay Region.* St. Petersburg, Florida.
33. *Tampa* (Fla.) *Tribune*, 29 Jan., 1991.
34. "The Lineup." *Tropical Fish Hobbyist* 39 (1991): 76-81.
35. "The Lineup." *Tropical Fish Hobbyist* 39 (1991): 138-141.
36. "The 23rd Environmental Quality Index." *National Wildlife* (1991): 33-40.
37. Vesilind, P. Arne and Peirce, J. Jeffrey. *Environmental Pollution and Control.* 2nd ed. Ann Arbor, Michigan: Ann Arbor Science Publishers, 1983.
38. "Walk a Mile for a Camel?" Worth Repeating. *Resources* 12 (1990): 16.